BASIC ELECTRONICS A

Introducing Electronics

Malcolm Plant

Hodder & Stoughton

LONDON SYDNEY AUCKLAND TORONTO

Better ask twice than to lose your way once.

Danish proverb

© 1990 SCDC Publications

First published in Great Britain 1976

Second edition 1990

British Library Cataloguing in Publication Data.

Plant, Malcolm
  Basic electronics. — 2nd ed.
  Bk.A. Introducing electronics
  1. Electronic engineering
  I. Title
  621.381

ISBN   0 340 41495 2

Typeset in 11/12 Baskerville by Taurus Graphics, Abingdon
Printed and bound in Great Britain for the educational publishing
division of Hodder and Stoughton, Mill Road, Dunton Green,
Sevenoaks, Kent by Thomson Litho Ltd, East Kilbride.

Basic Electronics is published in five parts

Book A Introducing Electronics
ISBN   0 340 41495 2

Book B Resistors, Capacitors, and Inductors
ISBN   0 340 41494 4

Book C Diodes and Transistors
ISBN   0 340 41493 6

Book D Analogue Systems
ISBN   0 340 41492 8

Book E Digital Systems
ISBN   0 340 41491 X

It is also available as one complete volume:
ISBN   0 340 41490 1

Note about the author
Malcolm Plant is a Principal Lecturer in the
Faculty of Education at Nottingham
Polytechnic. He is the author of several
books, including *Teach Yourself Electronics*
(Hodder & Stoughton 1988). His main
professional interests are in astronomy and
astrophysics, electronics instrumentation
and issues relating to conservation and the
environment.

Contents

Summary

This book begins by looking at ways in which electronics influences our daily lives – in entertainment, communications, control devices, and medicine for example. In Chapter 2, you will find a 'potted history' of electronics to help give a perspective to the rapid developments of electronics. Although many applications of electronics are far too complex to understand in detail, they can be regarded as a system of interconnected building blocks doing jobs such as counting, switching, amplifying, oscillating and so on. Some of these 'systems' concepts are introduced in Chapter 3.

Many of the very wide range of electronic functions rely on a few basic circuit ideas, i.e. the idea of current flow and control with switches, series and parallel circuits, and electrical energy. Simple experiments are included in Chapters 4 and 5 so that you can explore these ideas. A variety of circuit assembly methods is discussed in Chapter 6, which includes a list of the basic tools you require to assemble circuits. Chapter 7 looks briefly at atoms and ions to explain what electrons are and why they are so important to electronics.

The purpose of electronic circuits and systems is to produce electrical signals (e.g. a radio transmitter produces radio waves), or to change some property of electrical signals (e.g. an audio amplifier in a hi-fi system increases the power of audio signals). Electrical signals can be broadly divided into two types, direct current (d.c.) and alternating current (a.c.). Some of the basic properties, e.g. amplitude, of d.c. and a.c. signals are described in Chapter 8.

Chapter 9 describes the purpose of the multimeter which is the most useful test instrument for measuring voltage, current and resistance. The second most used instrument for test and measurement in electronics is the oscilloscope, and the ways of using this instrument are described in Chapter 10.

A third instrument, the waveform generator, which produces audio frequency sinusoidal and square wave signals is described in Chapter 11.

Chapter 12 is the first of five chapters (one at the end of each of Books A to E of *Basic Electronics*) which describe how to use a series of Project Modules. There are 35 of these Project Modules in *Basic Electronics* and the first seven of them are described in Chapter 12 of this book. They are built on printed circuit boards. The Project Modules provide a quick way of assembling useful circuits and examples are given of their applications.

As with every book of *Basic Electronics*, Book A ends with Answers to Questions in the text, Revision Questions and Revision Answers (Chapter 13).

If you wish to follow a quicker and easier route through Book A, you should omit the sections of the text marked with the symbols ∇ and $\triangle$ in the left hand margin.

1 Electronics Today

1.1 The electronic age

Electronics is about using things such as transistors and silicon chips to make electricity work for us. The first electronic device to be invented was the *valve* in the early years of this century, but not until the invention of the *transistor* in 1948 did electronics begin to have such a profound impact on the way we work, rest and play.

The transistor and the development of integrated circuits that followed have had far-reaching effects on nearly all aspects of life. Nowadays we take for granted the way electronics makes our lives comfortable, enjoyable and exciting. Let's begin by looking at five areas where it is having a big impact: consumer electronics, communications electronics, computer electronics, control electronics and medical electronics.

1.2 Consumer electronics

We are surrounded by products in and around our homes which make use of electronics in one way or another. It isn't always apparent that these products use electronic devices, but domestic equipment such as washing machines, central heating systems, cookers and burglar alarm systems do. And, in a more obvious way, so do a variety of home entertainment systems such as televisions, videos, hi-fis, electronic organs and home computers. For example, the *compact disc* has improved the quality of sound from a hi-fi system. This is a plastic disc about 120 mm in diameter and 1.2 mm thick (figure 1.1) which stores sound in the form of microscopically small pits, each about one thousandth of a millimetre long and about a ten thousandth of a millimetre deep in a continuous spiral track. The track is so fine that about thirty are as wide as a human hair. Sixty minutes of sound requires about 10 million such pits on one side of the disc.

The compact disc is coated with a layer of reflecting aluminium and covered by a protective film of transparent plastic. The disc is rotated at high speed in a player in which a finely focused beam of laser light 'reads' the information on the disc. The laser beam focuses so accurately on the surface of the disc that only the pits are read, not the dust or other irregularities on its surface. The sound produced by the compact disc is almost free from distortion e.g. surface hiss, and the system is compact and rugged enough for use by a jogger.

The camera is another example of a product that has been changed greatly by modern electronics. Miniature circuits on *silicon chips* in the camera enable perfect photographs to be taken without the user focusing or setting the exposure time (figure 1.2 overleaf). The circuits act as a tiny computer which works out the precise exposure time a picture requires

Figure 1.1 A portable compact disc and player
Courtesy: Morphy Richards CE Ltd.

and sets the shutter speed accordingly. On pressing the shutter release button, you may be sure the correct exposure will be obtained for the film you use.

Figure 1.2 The electronics inside a camera
Courtesy: Ferranti

1.3 Communication electronics

The achievements of modern communications are taken for granted. A well-established system now exists for world-wide radio and television communications using artificial satellites in orbit round the Earth, all part of a vast *telecommunications* system. For example, the Intelsat-5 series of satellites can each handle 12 000 telephone calls and two colour television channels and plans are underway to launch satellites weighing 50 tonnes capable of handling five million telephone calls and hundreds of television programmes. Some of these satellites are designed to beam television programmes direct to homes from a *geostationary orbit*. This is an orbit which keeps pace with the Earth's rotation so the satellite remains in a

Figure 1.3 Olympus, a new communications satellite Courtesy: British Aerospace

fixed position above a particular place on Earth. Figure 1.3 shows the first of the Olympus class of communications satellites which can beam television programmes and business information direct to aerials on rooftops, window sills and gardens. (figure 1.4).

The record for long distance radio communications is between spacecraft and their navigators on Earth. Packed with electronic equipment, the Voyager 2 spacecraft (figure 1.5) has completed its tour of the outer planets of the Solar System and has sent back a great deal of scientific information and detailed pictures about the planets Jupiter, Saturn, Uranus, and Neptune. Contact with Voyager 2 (and its predecessor Voyager 1 which has already left the Solar System) is expected to continue for some years, but it will take both spacecraft perhaps another 350 000 years to reach another star! Halley's comet was examined at close quarters by the Giotto spacecraft in 1986 (figure 1.6).

Figure 1.4 A dish aerial for receiving TV programmes from direct broadcast satellites (DBS) Courtesy: Science Photo Library/Adam Hart-Davies

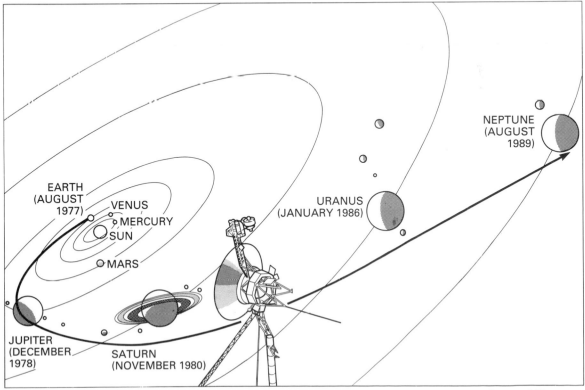

Figure 1.5 The spacecraft Voyager 2 has visited the planets Jupiter, Saturn and Uranus, and passed Neptune in 1989

Protected by a shield from collision with dust particles, Giotto used its dish aerial to send back colour pictures and measurements of the structure of this famous comet as it sped round the Sun before disappearing into the distant reaches of the Solar System for another 76 years.

In modern communication systems hair-thin *glass fibres* are being used instead of conventional wires (figure 1.7). Enormous amounts of information are being carried not by electricity but by pulses of laser light, reflected repeatedly from the inside walls of the fibre. There are some good reasons for developing *optical communications,* as this fast-growing system is called. Strong electric and magnetic fields, e.g. from lightning and electrical machinery, do not interfere with the message carried on the light beam, and broken fibres are not a fire hazard since the escaping light is harmless. Also glass fibres are cheaper and lighter than copper wires and can carry far more information. Optical fibres are particularly useful for replacing copper wires on aircraft and ships since fewer connections are required and because they weigh less.

1.4 Computer electronics

By comparison with the first valve computers of the 1940s, today's microcomputers show how dramatic the advances in electronics have been. Power-hungry, room-sized and unreliable, these early computers have been replaced by a variety of compact, efficient computers such as *calculators and microcomputers* (figure 1.8).

Such computers, as with so many other products and systems, owe their efficiency and compactness to the development of the integrated circuit on silicon chips (figure 1.9). This is a small piece of pure silicon on

Figure 1.6 Giotto, the spacecraft that flew through Halley's comet Courtesy: British Aerospace

Figure 1.7 Optical fibres   Courtesy: British Telecom

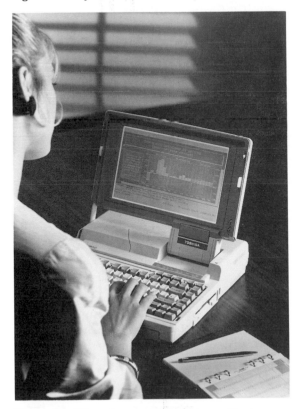

Figure 1.8 There is a trend towards smaller battery-operated computers   Courtesy: Toshiba

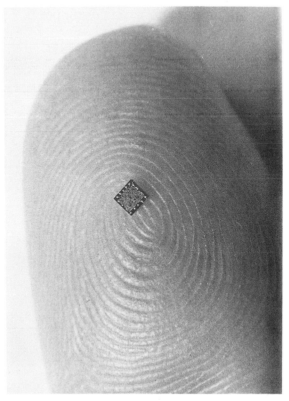

Figure 1.9 A silicon chip, unpackaged and useless on the author's finger

which complex miniaturised circuits are made by photographic and chemical processes. The manufacture and use of integrated circuits is called *microelectronics*.

The integrated circuit is cheap to make in large quantities, it can withstand rough treatment, it is reliable, it needs very little power to operate it and, above all, its small size enables it to be used in a wide range of products from wristwatches to spacecraft.

Integrated circuits are housed in plastic packages, like the one shown in figure 1.10. The microprocessor is undoubtedly the most complex and versatile of these integrated circuits. This device is the 'brain' of today's computers but, like the human brain, it has rather limited use on its own. Before it becomes a working computer it needs the support of a *memory* (another integrated circuit), a keyboard, a power supply and a display such as a television. However, the microprocessor controls all the operations of the computer and it does it fast. It is designed to work fast since a modern computer is called upon to process enormous quantities of information, as in a tax office, bank or police headquarters. To produce an accurate weather forecast, for example, even for two days ahead, requires a computer which can perform millions of calculations on information which is fed into it from aircraft, ships and weather balloons.

However, making fast work of calculations is not the only useful characteristic of a microprocessor. Because it can be programmed through a keyboard, or from an external memory, the basic microprocessor can be used as the 'brain' in a cash register, a washing machine, a juke box, or an industrial robot. So the fact that you can program a computer allows you to use the same machine to play a game of chess, or to find out what happens when a suspension bridge, say, is subjected to a strong wind.

1.5 Control electronics

Control is an essential part of the operation of any system. The safety of a car and its driver is dependent on many different types of control devices and systems. Electronic control mechanisms enable aircraft to land safely in poor visibility. Processes in a chemical factory depend on electronic control devices at various stages in the production of chemicals. The safe and reliable operation of power stations (figure 1.11) requires temperature and pressure to be under careful control. If something goes wrong, control systems should prevent serious damage by ensuring that the pressure does not build up and that the temperature is kept at a safe level.

There are many examples of electronic control systems in the home. Electrical power to lights, food mixers and drills is controlled by the flick of a switch. Temperature and humidity in the home or greenhouse can be controlled automatically. A washing machine operates under the control of a program which instructs control devices to carry out a preset sequence of washing activities, and so on.

When connected to electromechanical devices such as solenoids and motors, the computer can be used to control machines.

Figure 1.10 An integrated circuit, here a memory device for a computer, containing a silicon chip on which a complex circuit has been formed

Robots are computer-controlled machines. They are used for tasks which are repetitive (and often boring for humans) such as welding and the mounting of windows in a car manufacturing plant (figure 1.12 overleaf). In such places as nuclear reactors and interplanetary space, robots can work where it would be too hazardous for humans to do so.

Most robots use a microprocessor as a 'brain' to help them make decisions, e.g. to pick up a particular shaped object from a conveyor belt. Once programmed, the robot carries out this control job faultlessly. But suppose you want the robot to make decisions based on its accumulated experience? The robot is then said to have artificial intelligence (AI). Such an intelligent robot is being designed to explore the surface of Mars. It might have to decide whether to investigate yet another dried-up river bed, before or after making a detour to study a particularly interesting rock. The robot would need to take into account the length of daylight left on its part of Mars, on the time of the next pass of its parent satellite which is relaying signals to and from Earth, and on many other factors as well.

Present-day robots are not up to the job of complex decision-making of this kind. The microprocessors they use are too slow, they cannot handle many pieces of information simultaneously. What is more, you cannot talk to today's robot. Robots of the future will be using a revolutionary type of microprocessor called a transputer. A transputer is designed to 'talk' to other transputers; instead of one transputer, future robots will be using many interconnected transputers so that together they are able to make rapid decisions. The word comes from a contraction of 'transistor' and 'computer' implying that transputers are intended to be used in future computers just as freely as transistors have been used and have improved present-day computers. We shall soon be using these 'supercomputers' in robots capable of understanding spoken

Figure 1.11 A central control room of the Magnox power station on the northern shore of Trawsfynydd Lake in Clwyd. This room contains the main switchgear, reactor and turbogenerator controls. From this room it is possible to control all essential operations of the power station Courtesy: Central Electricity Generating Board

Figure 1.12 A robot fits a rear screen to a Montego car at the Cowley assembly plant
Courtesy: Austin Rover Group Ltd.

commands, and of helping doctors, engineers and scientists to make decisions rapidly (see also section 2.7).

1.6   **Caring electronics**

Electronics has helped greatly in diagnosing people's health problems and in performing the operations to cure them. An *electrocardiograph* is used for diagnosing heart defects by recording the electrical signals produced by the heart using electrodes attached to the chest. In a hospital, an electrocardiograph would be part of a complete patient-monitoring system which includes instruments for measuring respiration rate, blood pressure and body temperature. But if you just want to check how fast your heart is beating during exercise, the portable electronic instrument in figure 1.13a could be used.

Another diagnostic medical instrument of great use in hospitals is the *ultrasonic scanner* (figure 1.14) which provides pictures

Figure 1.13a An instrument for measuring heart pulse rate
Courtesy: Centronic Health Care Services

Figure 1.13b Monitoring your heart beat wasn't much fun in the old days! Courtesy: Picker International Ltd.

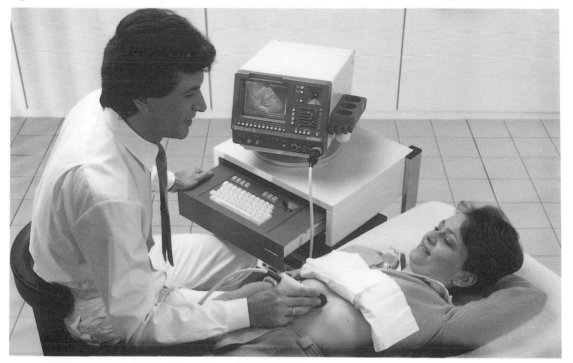

Figure 1.14 An ultrasonic scanner Courtesy: Siemans

of the inside of the body using high-frequency sound waves. A device which generates these waves is placed on the skin and the echo pattern from these waves bouncing off organs in the body, is processed electronically using a computer to give an image on a screen. The ultrasonic scanner is considered to be much safer than X-rays, especially when examining a developing baby in the womb.

Microelectronic devices are so small and require so little power to operate that they are being considered for direct implanting in the body. Research is being directed towards a video camera small enough for putting in the eyesocket of blind people. The electrical image generated would stimulate the brain's seeing area — the visual cortex. The idea of a silicon chip which responds to the chemicals surrounding living tissue is also being pursued. Such a *biochip* could be implanted in the heart of a person using a pacemaker. It would respond to hormones in the blood and adjust the pulse rate of the pacemaker to make the heart beat faster or slower just like a normal heart.

Questions

1 Which of the following statements is true about a compact disk:
(a) it is a type of computer memory;
(b) it stores music and speech on its surface in the form of millions of microscopic pits;
(c) it is a smaller version of the common vinyl disk.
2 Explain what is meant if a camera is said to have 'automatic focusing'.
3 What is a silicon chip?
4 What can robots do more easily than us?

5 State three ways a microcomputer is useful in the home.
6 Give two advantages of using weather satellites to forecast the weather.
7 Give three examples of how electronics helps blind people to lead a fuller life.
8 Which one of the following statements is true?
A transputer is:
(a) a type of microcomputer;
(b) a new way of moving people around;
(c) a revolutionary type of microprocessor.
9 An electrocardiograph is:
(a) a type of electronic game;
(b) an instrument for showing the shape of electrical signals from the heart;
(c) a device for stimulating the muscles.
10 A satellite is said to have a geostationary orbit when it
(a) stops the Earth rotating;
(b) remains over one part of the Earth's surface;
(c) is seen from all parts of the Earth's surface.
11 Discuss in a few words the worry that people have about storing personal information on computers.
12 Discuss in a few words the advantages and disadvantages of putting cars under automatic control on motorways.
13 A wristwatch-sized 'communicator' enabling us to see and talk to another person anywhere in the world may be available in the future. Would you welcome such a device or would you have some reservations about using it?

2 A Short History of Electronics

2.1 Introduction

Less than a hundred years ago, electronics was unknown. There were no radios, televisions, computers, robots, satellites, none of the electronic products we now take for granted. In such a short time there has been a revolution in the ways we communicate, control, measure, and entertain ourselves. This revolution came naturally out of the study of electricity, an old science not unknown to the Greeks over 2000 years ago. Electricity was a subject of great interest to Victorian scientists, and to Sir William Crookes in particular.

2.2 The discovery of cathode rays

The beginnings of electronics can be traced back to the discovery of *cathode rays* in the closing years of the last century. These mysterious rays had been seen when an electrical discharge took place between two *electrodes* in a glass tube from which most of the air had been removed. Sir William Crookes called these rays 'cathode rays' since they seemed to start at the negative electrode (the cathode) and moved towards the positive electrode (the anode).

At that time nobody had any idea what cathode rays really were. But during an historic lecture at the Royal Institution in London in April 1897, Sir J.J. Thomson declared that cathode rays were actually small, rapidly moving electrical charges. Later these charges were called *electrons* after the Greek work for *amber*.

Amber is the fossilized resin from trees and has peculiar properties, as the ancient Greeks had found. If rubbed with fur or a dry cloth, it has the power to attract small pieces of dust and fluff. Neither these Greeks, nor the scientists who devoted so much time to studying its properties in the period from the 17th century, had a successful explanation of why amber behaved in this way. But the discovery of the electron provided the answer. We now know that the electrical behaviour of amber (and of many other electrical insulators) is caused by *static electricity*. Friction between the cloth and amber causes electrons to be transferred from the cloth to the amber where they remain and give amber an overall negative charge. This charge causes it to attract small bits of material to it. Vinyl discs we play on a hi-fi show this effect, too.

2.3 The invention of the valve

The first practical application of cathode rays was the invention of the *thermionic valve* by Sir John A. Fleming in 1904. In this device, electrons were produced by heating a wire (the filament) in an evacuated glass bulb. The word 'thermionic' comes from 'therm' meaning heat, and 'ion' meaning a charged particle, i.e. the electron. Electrons ejected from the heated filament (the cathode) moved rapidly to an anode provided the anode was at a more positive voltage than the cathode. The flow of electrons stopped if the anode's voltage was made less than the cathode's. This electronic component was called a *diode*

Figure 2.1 A 1920s Echodyne superhet radio receiver Courtesy: Science Museum

since it had two electrodes. It acted like a valve because electrons flowed through it in only one direction, from the cathode to the anode, not in the opposite direction.

It did not take long for an American, Lee de Forest, to make a much more interesting and useful thermionic valve. By putting a third electrode, made of a mesh of fine wire through which the electrons could pass, between the anode and cathode of the diode he produced a *triode*. By adjusting the voltage on the middle electrode (called the *grid*), he was able to make the triode behave like a switch or, more important, as an *amplifier* of weak signals. The triode made it possible to communicate over long distance by radio. The importance of this development was demonstrated dramatically in 1912 when the luxury liner *Titanic* collided with an iceberg in the

Atlantic Ocean. As this 'unsinkable' ship was going down, her radio operator broadcast an SOS radio signal, using Morse code, which was picked up by ships in the area.

2.4 The beginnings of radio and television

Strangely, the First World War (1914–18) did little to stimulate applications for thermionic valves. But immediately after the war, the demands of entertainment gave electronics a push which has gained strength ever since. In London the British Broadcasting Corporation was formed, and in 1922 its transmitter (called 'call sign 2LO') went on the air. Firms such as Marconi, HMV and Echo made radio sets from components and valves supplied by Mazda, Ozram, Brimar and others.

The second major boost to this emerging electronics industry was the start of regular television transmissions from Alexandra Palace in London in 1936. But at that time, the public had little interest in television, which was hardly surprising as the pictures produced by John Logie Baird's mechanical scanning system were not very clear. By the time EMI had developed an electronic scanning system which gave much better pictures, the Second World War had begun. The Alexandra Palace transmitter was closed down abruptly in September 1939 at the end of a Micky Mouse film. Britain feared that the Germans might use the transmissions as a homing beacon for its aircraft to bomb London.

2.5 Radar and the Second World War

However, from 1939 to 1945, there were important advances in electronics. Perhaps the most significant invention was *radar* — developed in Britain to locate enemy aircraft and ships.

The word radar is an acronym for it is formed from the words *r*adio *d*etection *a*nd *r*angefinding. Radar was made possible by the invention of a high-power thermionic valve called a *magnetron*. This device produced high-frequency pulses of radio energy which were reflected back from an aircraft or ship to reveal their range and bearing. Magnetrons are the source of microwaves in microwave cookers.

Figure 2.3 An early radar receiver, *c*. 1940
Courtesy: Science Museum

Figure 2.2 A 1936 Marconiphone television and radio receiver   Courtesy: Science Museum

Figure 2.4 Three generations of electronic components; the valve, the transistor and the integrated circuit

2.6 The invention of the transistor

In the period immediately following the Second World War there was a major step forward in electronics brought about by the invention of the first working *transistor*. In 1948, Shockley and his co-workers, Bardeen and Brattain in the Bell Telephone Laboratories in the USA, demonstrated that a transistor could amplify and act like a switch. Actually the idea of using the element *germanium* to produce the solid-state equivalent of the vacuum tube triode had been worked out about 25 years earlier. However, the way current flowed in semiconductors, as these germanium-based transistor materials were called, was not very well understood. Furthermore, until the 1950s it was not possible to produce germanium with the very high purity required to make useful transistors.

These transistors turned out to be very successful rivals to the thermionic valve. They were cheaper to make since their manufacture could be automated. They were smaller, more rugged and had a longer life than valves, and they required less power to operate them. Once *silicon* began to replace germanium as the basic semiconductor for making transistors in the 1960s, it was clear that valves could never compete with them for reliability and compactness. Consider ENIAC (another acronym meaning *e*lectronic *n*umerical *i*ntegrator *a*nd *c*alculator). This was a *computer* which used valves and was built at the University of Pennsylvania in the 1940s. ENIAC filled a room, was worked by 18 000 valves, needed 200 kW of electrical power, had a mass of 30 tonnes and cost a million dollars. The transistorised desk-top *calculator* of the 1960s was battery-powered, had a mass of a few kilograms and was capable of far more sophisticated calculations than ENIAC. This trend towards low-cost yet more complex circuits, to greater reliability yet lower power consumption, continues to be an important characteristic of the development of electronics.

ENIAC is now regarded as a *first generation computer*, and the transistorised computers which followed it in the 1960s as *second generation computers*.

2.7 Silicon chips make an impact

During the 1960s, the first integrated circuits were made. Techniques were developed for forming up to a few hundred transistors on a single *silicon chip* and linking them together to produce a working circuit. Computers using this new silicon chip technology are now known as *third generation computers*. The Apollo spacecraft which took man to the Moon in the late 1960s and early 1970s used these computers for navigation and control.

The art of putting complex circuits on silicon chips a few millimetres square is called *microelectronics*, and the devices produced began to have a significant impact on work and leisure activities through the 1970s. The stimulus to miniaturize circuits came from three main areas: weapons technology, the 'space race' and commercial activity.

Modern weapons systems depend for their success on circuits which are small, light, quick to respond, absolutely reliable, and which use hardly any electrical power. Miniature circuits on silicon chips offer these advantages. The 'space race' began when Russia launched Sputnik in 1957. At first, America's response was unsatisfactory, but she gained ground during the 1960s and put men on the Moon at the end of the decade. Lacking the enormously powerful booster rockets developed by Russia, America needed compact and complex spacecraft which stimulated the design of small and reliable control, communications and computer equipment. During the 1970s, spin-offs from military interests and the space race stimulated the growth of an electronics industry bent first on creating electronic goods and then satisfying the demand for them in the home, the office and industry.

During the 1970s, the number of transistors integrated on a silicon chip doubled every year and this trend continues. Along with this increasing circuit complexity has been a similar doubling in the information processing power of the silicon chip. The most important of these chips is the microprocessor; it contains most of the components needed to operate as the central processor unit of a computer. It is a highly complex device which can be programmed to do a variety of tasks. This versatility means that it acts as the 'brain' in a wide variety of devices. These *fourth generation computers* have become faster and cheaper; they are now used in industrial robots and sewing machines, in the Space Shuttle and toasters, in medical equipment and computer games. Their programmability and cheapness are their strengths. In fact they are often the cheapest component in an electronic system.

The microprocessor brings the story to the present day. The highlight of present-day microelectronics is the *transputer* (see also section 1.5). This is a revolutionary microprocessor which has been developed in Britain by INMOS for the fifth-generation computers. The transputer's name is derived from '*trans*istor' and 'com*puter*' which underlines the idea of an innovative microprocessor which will become a universal building block used with others in large quantities just as transistors are today. A typical transputer like the one shown in figure 2.5 contains a

Figure 2.5 The transputer Courtesy: INMOS Ltd.

quarter of a million transistors on a 7mm by 7mm silicon chip. It is capable of performing ten million instructions per second. The chip contains special circuits to enable it to communicate with other transputers to increase the overall computing power. Such an array of transputers will enable supercomputers to be built that are capable of intelligent interaction between people and machines. People will be able to hold a conversation between a supercomputer (or *fifth generation computer*) and consult it as an expert in a chosen field of knowledge. The transputer will make possible intelligent robots that can see and hear spoken commands. They will be involved in decision-making in industry, government and the home.

3 Electronic Systems

3.1 A transport system

The word 'system' is used a lot in everyday conversation. We talk of a telephone system, an education system, a music system, and so on.

A familiar transport system is shown in figure 3.1. It helps people to find their way about a city by bus. A map such as this one isn't any good at telling us how long the journey takes, or what there is to see on the journey, or what colour the buses are painted. All it tells us are the names of places linked together by the various Cityline bus routes. That's the purpose of a systems diagram such as this one: it gives us essential information about what the system provides without confusing us with unnecessary detail.

This way of looking at a complex activity and simplifying it by means of a systems diagram is very useful when dealing with the applications of electronics.

3.2 Black boxes

An electronic circuit is designed to do something useful such as amplifying, switching, sensing, counting or displaying. In order to understand the overall purpose of the circuit it isn't necessary to know *how* the amplifying or counting part works, just *what* it does. So we draw the amplifier or counter as a box as shown in figure 3.2. This box is called a *black box* since we are interested only in what it does, not what goes on inside it. The box has information going into it called the input, which it acts upon in some way to produce the required output information. In electronics, a black box is an *activity box* so what comes out of it is different from what goes into it.

A *record player* can be looked at as a collection of black boxes making up a block diagram as shown in figure 3.3, overleaf. Of course, such a system has many more parts than a *pickup, amplifier* and *loudspeaker*, but

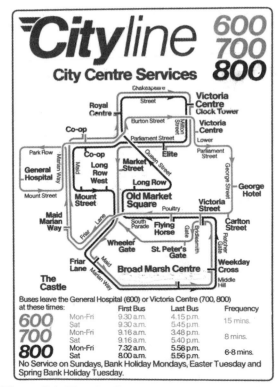

Figure 3.1 The 'Cityline' bus route map
Courtesy: Nottingham City Transport Limited

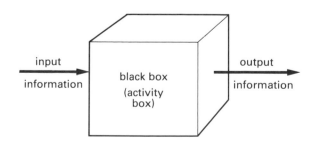

Figure 3.2 A black box circuit building block

these are its main activity boxes. The amplifier takes the signal from the pick-up, gives it more power and sends it off to the loudspeaker. Thus the action of this black box is amplification. The pick-up converts the mechanical movement of the stylus into an electrical signal suitable for the amplifier to work on. And the loudspeaker has the job of producing soundwaves from the signals provided by the amplifier. It is then a biological system, our ears, that converts the sound waves to the sensation of hearing.

3.3   Examples of electronic systems

When designing and making electronic circuits, you should try to describe the purpose of the circuit by drawing it as a block diagram. A block diagram will help you to understand *what* part each black box plays in the system even though you may not be sure *how* it achieves what it does.

Many of the circuits in *Basic Electronics* are

to do with communications, control, computing and instrumentation. You can get a good idea about how these circuits work by regarding each one as a system made up of black boxes.

Figure 3.4 shows a block diagram of a *communication system*. A *transmitter* sends the information along a *communication channel* at the end of which is a *receiver*. In a simple home intercom, for example, the transmitter would be a microphone coupled to a loudspeaker acting as the receiver. The communications channel would be the wires which connect the transmitter to the receiver. Of course, the telecommunications system which allows you to talk to someone in Australia is a much more complex affair than the home intercom but the main features of it are the same as shown in figure 3.4.

A block diagram of a *computer system* is shown in figure 3.5 and is built around a microprocessor which is sometimes called a CPU or *central processor unit*. The CPU carries out a list of instructions (a *program*) which is held in the *memory*. The *input ports*

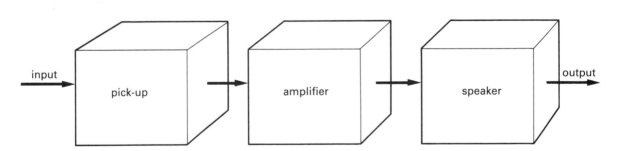

Figure 3.3 A block diagram of a record player

Figure 3.4 A communications system

and *output ports* are the computer's 'windows' to the outside world through which information comes and goes. Within the computer, information flows along the computer highway, or *bus,* which connects together the various parts of the computer system.

Figure 3.6 shows the main parts of a *control system.* The input information is the value of what it is you want to control. For example, if you heat a tropical fish tank with an electrical heater, the input information is the water temperature. A *comparator* compares the value of the temperature you want the water to be with information about the actual temperature which flows through the feedback path. If the temperature is too high, the *driver stage* is instructed to switch off the heater: if too low, to switch it on. Thus the temperature of the tank is kept near to the temperature you want. This control system applies equally well to mechanical and biological

systems. For example, when your body temperature is too high, you sweat to cool down. If it is too low you shiver to warm up.

The *instrumentation system* shown in figure 3.7 (overleaf) has three main building blocks. The *sensor* converts the quantity you want to measure into an electrical signal. A *signal processor* converts this signal into a form which is suitable for operating the *display.* The sensor of a thermometer, for example, would change the effect of heat into an electrical signal. How this signal is processed depends largely on how the temperature is to be displayed. If the temperature is to appear as a reading on a scale over which a pointer moves, amplification of the signal from the sensor is generally sufficient. If a display of numbers is required, more complex signal processing is necessary. These two main methods for processing information are discussed in Section 3.5.

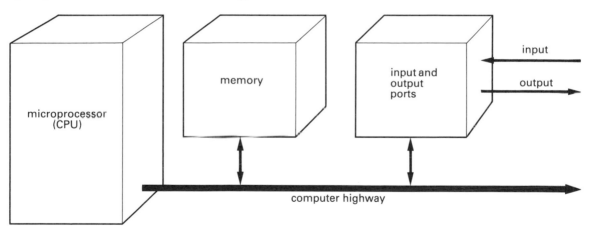

Figure 3.5 A computer system

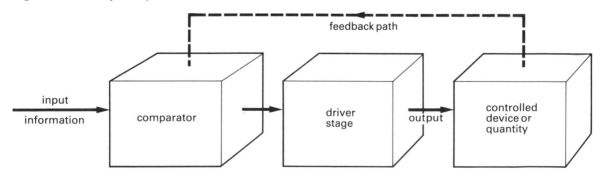

Figure 3.6 A control system

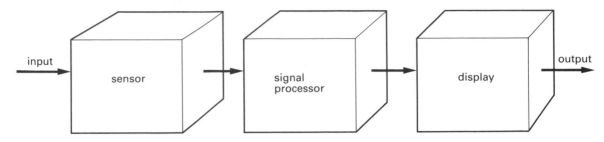

Figure 3.7 An instrumentation system

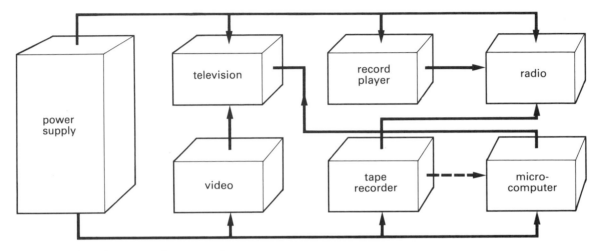

Figure 3.8 The subsystems of an entertainment system

3.4   **Systems and subsystems**

A block diagram of an electronic system shows as much or as little detail as we like. Let's use the example of the record player (figure 3.3) to show how this detail can vary.

Figure 3.8 shows the block diagram you might draw if you were planning to install an electronic entertainment system in your home. The record player is now just one part of this larger system and is drawn as a single black box. The diagram shows how the power supply connects to each part of the entertainment system. The dotted lines indicate connections you might want to make. For example, you could use the cassette recorder to store a program for the microcomputer.

Each black box in the entertainment system is known as a *subsystem*. So whereas

figure 3.3 shows the amplifier as a subsystem of the record player, the record player is now a subsystem of the entertainment system. Of course, you could think of the amplifier as a system made up from three subsystems: pre-amplifier, tone control and power amplifier. And if you wanted to show more detail still, the tone control, for example, could be regarded as a set of subsystems. The individual components which go to make up the tone control circuits, would be regarded as the black boxes in this system.

Microelectronic devices in integrated circuit form are often complete subsystems which require connection to input and output devices and a power supply to make a complete system. Amplifiers, timers, calculators, radios and microprocessors are now made as complete subsystems on silicon chips.

3.5 Analogue and digital systems

The words *analogue* and *digital* are in general use for describing features of electronic systems. For example, watches and clocks use analogue or digital methods for displaying time — see figure 3.9.

The word analogue means 'model of', and the analogue watch models the smooth passing of time by using hands which move smoothly over its face. The word digital means 'by numbers' and the digital watch displays the passing of time by means of numbers which change every hour, minute or second. Note that the analogue display shown in figure 3.9 relies on digital circuits for keeping time — only the display is analogue.

It is quite common for electronic systems to combine analogue and digital electronic circuits, but, first, let's look at two systems, one entirely analogue in operation and the other entirely digital. The photographic exposure meter shown in figure 3.10 is a good example of an entirely analogue system. The smooth variation in daylight from dawn to dusk, or from outdoors to indoors, is converted by a light sensor into an electrical analogue, or model, of this variation. This electrical analogue is then passed to an information processor, such as an amplifier, before operating a pointer moving over a scale. Figure 3.11 is a block diagram of this exposure meter.

Figure 3.9 Analogue and digital displays on watches
Courtesy: Seiko

Figure 3.10 An analogue photographic meter
Courtesy: Weston Instruments

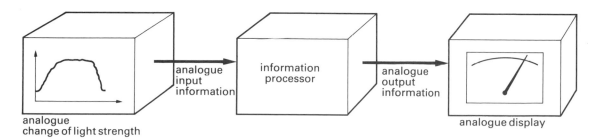

Figure 3.11 An entirely analogue electronic system

Figure 3.12 A digital tachometer Courtesy: Electroplan Ltd.

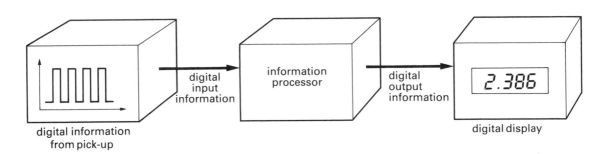

Figure 3.13 An entirely digital electronic system

The tachometer shown in figure 3.12 is an entirely digital system. This instrument is used for measuring frequency of rotation of, say, an axle or wheel. It gives a digital readout of the frequency of rotation measured in revolutions per minute. Figure 3.13 shows that the input information is made up of a series of *pulses* which are produced by the pick-up as it senses the rotation of the wheel. The information processor in this case is a counter which counts the number of pulses in a period of time and sends this information to the digital display.

The digital thermometer shown in figure 3.14 is an example of a digital system which has an analogue input and a digital output. Figure 3.15 is the block diagram for this thermometer. A sensor converts the analogue changes of temperature into an equivalent electrical analogue of this change. A building block called an *analogue-to-digital converter (ADC)* converts this analogue information into its digital equivalent which is then in a suitable form for operating the digital display. The way an analogue-to-digital converter works is explained in Chapter 10 of Book D.

The ADC is a very important building block in modern circuit and system design for two reasons. First, digital displays are becoming commonplace on all sorts of equipment. Second, since computers handle digital not analogue information, any information fed to them which is in analogue form, such as temperature,

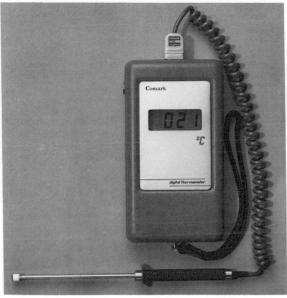

Figure 3.14 A digital thermometer Courtesy: Comark Electronics Ltd.

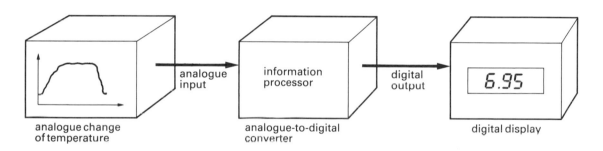

analogue change of temperature

analogue-to-digital converter

digital display

Figure 3.15 The system for a digital thermometer

pressure and distance, must be converted into a form which the computer will recognise. As you might expect, the reverse process is necessary if you want to use a computer to control the brightness of a lamp or the speed of a motor. In this case, a digital-to-analogue converter (DAC) is used. (Chapter 10 of Book D explains how a digital-to-analogue converter works.)

Questions

1 Which one of the following things are all electronic systems designed to do?
   (a) Reduce the cost of doing something.
   (b) Act on information in a useful way.
   (c) Play games.

2 What is (a) the input and (b) the output of a central heating system?

3 What is a 'black box' in electronic circuits? Why is it useful to look upon an integrated circuit as a black box?

4 What is the name of (a) the input device (b) the output device of a record player?

5 Name three electrical functions that are subsystems of a car.

6 Which of the following quantities are in analogue form: (a) speed of the wind (b) the size of a crowd of people (c) the length of a day (d) the temperature of air?

4 Simple Circuits

4.1 Components, circuits and symbols

The circuit designer connects components together with the aim of producing a circuit or system which does something useful. Transistors, resistors, capacitors, diodes, lamps and switches are examples of *discrete* components. A microprocessor is an example of an *integrated* component.

A circuit diagram shows how components must be connected together to produce a working circuit. Each component is drawn as a symbol as the examples in figure 4.1 show. These symbols are those recommended by the British Standards Institute in its document BS3939 *Graphical Symbols for Electrical Power, Telecommunications and Electronics Diagrams*.

Figure 4.2 shows the circuit diagram of an audio amplifier such as might be used in a record player system. This circuit uses an integrated circuit and three different types of discrete components. Can you identify them from figure 4.1?

It is most important that a circuit diagram shows connections between components clearly. Not only should the correct symbols be used, but interconnections between the *conductors*

COMPONENT	SYMBOL	DESCRIPTION
		RESISTOR, FIXED VALUE. Values from less than one ohm to many millions of ohms. Different materials and power ratings – see Book B. Used to control currents and hence voltages.
		RESISTOR, VARIABLE VALUE. Adjustable by knob on the spindle. Adjusts from zero ohms to maximum of device. Can be used as a potentiometer – see Book B.
		CAPACITOR, FIXED VALUE. Values from farads to billionths of a farad. Type shown made from polyester but many other types – see Book B.
		CAPACITOR, FIXED VALUE. But this one is an electrolytic type. Also polarised so must be connected correctly in DC circuits. Type above is unpolarised. Both types store electric charge – further details in Book B.
		DIODE. Electronic valve. Current flows easily in direction of arrow, with extreme difficulty in opposite direction – details Book C.
	(a) (b)	TRANSISTOR. Two main types, (a) bipolar, (b) unipolar or field-effect transistor. Used for amplification and switching and the basic building block of integrated circuits. Both transistors used in projects (chapter A6) and further details in Book C.
		LIGHT DEPENDENT RESISTOR. Very useful device for converting light changes into electrical changes. See Book B for more information.
		LIGHT-EMITTING DIODE. Emits light when current flows in direction of arrow. Red, green, yellow and blue types available.
		SEVEN-SEGMENT DISPLAY. Made from light emitting diodes to enable ten digits and some letters to be displayed. A decimal point is provided. Available as single displays or/and as groups of 2, 4, etc. Gives bright display but liquid crystal types use less power so better for portable projects – see Book E.
		INTEGRATED CIRCUIT. Many thousands of different ICs from simple amplifier shown to microprocessor. Used in practical circuits in this book, for example in Chapter A6.
		SWITCHES. Mechanical push-switch shown. Used widely for on-off control of power to a circuit but also as 'starters' and 'stoppers' in counting circuits. Further types and applications in Chapter A5.

Figure 4.1 Some electronic components and their symbols

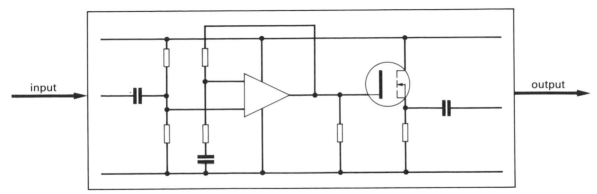

Figure 4.2 A circuit diagram of an amplifier

(wires) should be drawn horizontally and vertically except in special cases where they may be at an angle to the horizontal. Figure 4.3 shows the rules for conductors which meet in circuit diagrams. Wherever possible, circuits and systems diagrams should be drawn so that inputs to the circuit are on the left and outputs on the right as shown in figure 4.2 and similarly for the block diagrams described in chapter 3.

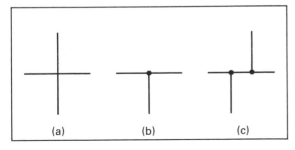

Figure 4.3 How to show connections between conductors in circuit diagrams: (a) no connections; (b) single connection; (c) double connection

The flow of electrons is called an *electric current* and is measured in units of *amperes*.

Figure 4.4 shows a simple circuit in which a battery provides an electromotive force of 6 volts (6 V). The current which flows in this circuit has a value of 0.06 amperes (0.06 A). Note that the battery symbol is made up of four *cell* symbols. A cell, such as one you put into your torch, usually has an e.m.f. of 1.5 V. By connecting four cells end-to-end as shown in figure 4.4, an e.m.f. of 6 V is obtained.

Electrons flow from the negative terminal of the battery to the positive terminal, continuously passing through the lamp on the way. This flow of electrons is known as an *electron current*. Before electrons were discovered, electricity was thought to flow from the positive terminal to the negative terminal of a battery. This direction is still used in circuit diagrams and is called *conventional current*. It usually doesn't matter which direction current flow is drawn, provided you stick to one or the other.

4.2 **Making electrons move**

Just as a mechanical force is required to make a car move, an electrical force is required to make electrons move. This electrical force is called an *electromotive force* (abbreviated to e.m.f.) and is measured in units of *volts*. A battery produces an e.m.f. which will make electrons move in a circuit.

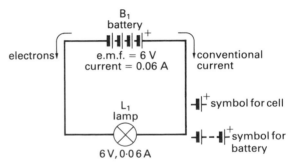

Figure 4.4 A simple circuit diagram

The electric current flows easily through the connecting wires, which are called conductors, since they are made of copper which has a small electrical *resistance*. However, the filament of the lamp is made of the metal tungsten which has a much larger resistance than copper. In passing through the filament, the electrons do far more work than when they pass through the copper wire and heat is produced. This heat raises the temperature of the filament to about 1000°C so that it becomes white hot and emits light. The unit of electrical resistance is the *ohm* and you will see how it is related to the ampere and the volt in Book B.

4.3 Series and parallel circuits

Figure 4.5 shows two lamps, L_1 and L_2, connected in *series*. It is obvious that whatever current flows through L_1 must also flow through L_2. Thus 'series' means one after the other. In this circuit there is only one path for the current, through each lamp in turn.

Figure 4.6 shows two lamps connected in *parallel*. The current which flows from the

battery divides. Part of it flows through L_1 and part through L_2. Thus 'parallel' means side-by-side. In this circuit there are two paths for the current to flow through.

In every circuit you design and build there are components connected in series or in parallel with each other. Even quite simple circuits have components connected in a combination of series and parallel arrangements. For example, figure 4.7 shows lamp L_1 connected in series with two parallel-connected lamps, L_2 and L_3.

4.4 Putting simple circuits together

Circuits which are made from just a few discrete components can be put together on a length of *screw terminal block*. Three lengths of this block can be seen in figure 4.8. The '12-way' length of screw terminal block in the figure has twelve short brass tubes mounted in a plastic strip. Each tube carries two screws which are tightened using a small screwdriver to clamp wires which are put into the tube. This assembly method is cheap and quick and will be recommended for some of the circuits described in this book. More complex circuits, and those using integrated circuits, require a special assembly-system called a *breadboard* and this is introduced in Section 6.1.

Figure 4.8 also shows some of the other components and materials required for the simple experiments described below. The power supply is a 6 V battery made from

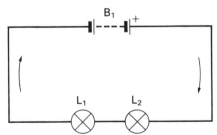

Figure 4.5 Lamps connected in series

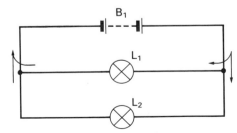

Figure 4.6 Lamps connected in parallel

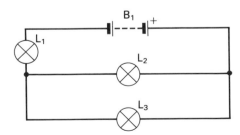

Figure 4.7 Lamps connected in series-parallel

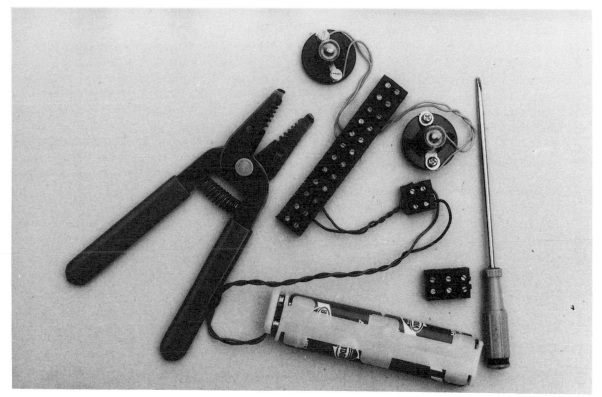

Figure 4.8 Materials for building simple circuits

four 1.5 V cells connected in series in a *plastic cell holder*. The power supply is connected to the circuit using the *battery clip* which has two wires each ending in about 10 mm of bare wire. Short lengths of *link wire* are necessary in some circuits. This wire should be plastic-covered, single-strand wire, 0.6 mm in diameter. The ends of the link wires should be stripped for about 10 mm using *wire strippers*.

4.5 *Experiment* A1

Making a simple series circuit

Use the screw terminal block to connect two lamps in series as shown in figure 4.9. When you have completed the connections, see if you can answer the following questions, overleaf.

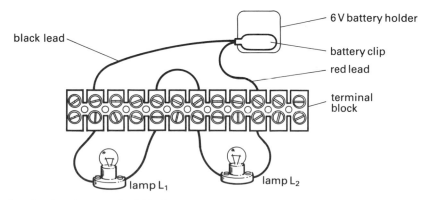

Figure 4.9 A series circuit using terminal block

1   Compare the brightness of one lamp with that of the other. Do you think that the same current is flowing through each lamp?
2   Unscrew one lamp. Why do both lamps go out?

It is important for you to notice that if the current is prevented from flowing through one lamp, it is also prevented from flowing through the other lamps in series with it.

3   Connect a wire link across lamp L_1. Why does this lamp go out? Why has lamp L_2 become brighter?

1   Unscrew one lamp. Why does the other lamp remain lit?
2   When both lamps are in circuit, why are they equally bright?
3   Why doesn't the brightness of the second lamp change when the other lamp is unscrewed?

Be sure you understand that in a parallel circuit if you break the current flow through one lamp, the other lamp does not go out since it is still connected across the battery terminal.

4.6 *Experiment* A2

Making a simple parallel circuit

Use the screw terminal block to connect two lamps in parallel as shown in figure 4.10. When you have completed the connections, see if you can answer the following questions.

4.7 *Experiment* A3

Making a simple series–parallel circuit

Use the terminal block to wire up the circuit shown in figure 4.7, then answer the following questions.
1   Explain what happens when a wire link is connected across lamp L_1. What sort of circuit do you have now?
2   Explain the effect of connecting a link across lamp L_3.

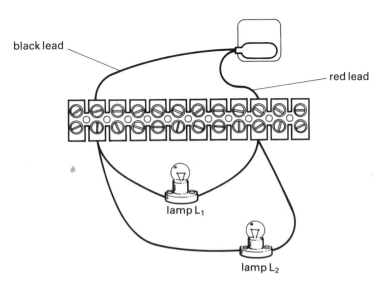

black lead

red lead

lamp L_1

lamp L_2

Figure 4.10 A parallel circuit using terminal block

5 Using Switches

5.1 The importance of switches

A switch is used to control the flow of electric current through a circuit. In the simple circuit shown in figure 5.1, a simple on-off switch is shown. If the switch is *open*, a current cannot flow through the lamp since the conducting path is broken. If the switch is *closed*, a current can flow since a conducting path is made.

We use simple on-off switches like this dozens of times every day. If you drive a car or motor bike, you will operate the ignition, lights, heater and radio using on-off switches. Cookers, fridges, hi-fi systems, television, radio, lights and heaters are just a few of the appliances in the home which are operated by on-off switches. Even the keyboard switches of calculators, computers and electronic games are simple on-off switches. Mechanical switches of various designs are generally used for on-off control of electricity and a few of these are discussed in this section. Later you will use non-mechanical switches which are made from devices such as the transistor. These switches have no moving parts. Digital electronic devices such as computers and calculators use thousands of transistor switches which are switched on and off very fast when calculations are being made.

5.2 Types of mechanical switch

Many switches require a mechanical force to operate them. The force brings together, or separates, electrically conducting metal contacts.

(a) Push Button switch The switch shown in figure 5.2 (overleaf) is a simple 'push-to-make, release-to-break type'. There are two symbols for this type of switch depending upon whether pushing 'makes' or 'breaks' the contact — see figure 5.3 overleaf. A simple on-off switch for a table lamp, for example, is a 'push-to-make, push-again-to-break' switch.
Are there any push-button switches in your home?

(b) Slide switch and toggle switch Figure 5.4 (overleaf) shows the appearance of two common types of these switches. They are manufactured as either *single-pole double-throw* (s.p.d.t.) or *double-pole double-throw* (d.p.d.t.). The circuit symbols for these types are shown in figure 5.5, overleaf. The poles of the switch are the number of separate circuits the switch will make or break simultaneously. Thus the d.p.d.t. type will operate two circuits at the same time.

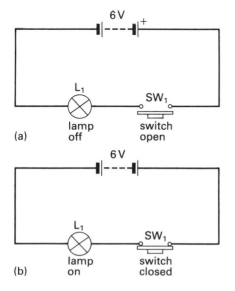

Figure 5.1 (a) Switch open, lamp off (b) switch closed, lamp on

The s.p.d.t. type is sometimes knows as a *change-over switch* since the pair of contacts which is 'made' changes over as the switch is operated. One common use for an s.p.d.t. switch is to operate a light from two positions, the top and bottom of a stairs, for instance — see Section 5.5.

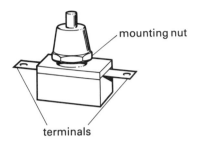

Figure 5.2 An example of a push-button switch

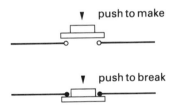

Figure 5.3 Circuit symbols for two types of push switch

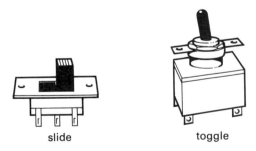

Figure 5.4 Examples of slide and toggle switch

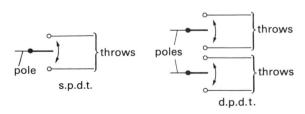

Figure 5.5 S.p.d.t. and d.p.d.t. switch symbols

(c) Microswitch There are two main features of this switch: a small force is required to operate it, and this force need act through only a small distance. In figure 5.6 the microswitch is fitted with a lever making the force required even less, although this does mean that the force has to act through a larger distance.

(d) Rotary switch In the switch shown in figure 5.7, one or more fixed contacts (the poles) make contact with switch contacts mounted on a spindle. Thus a number of separate switching combinations can be made. 1 pole, 12 way; 2 pole, 6 way; 3 pole, 4 way; 4 pole, 3 way and 6 pole, 2 way are some of the switching combinations for a rotary switch of this type.

(e) Electromagnetic relay When a small current is passed through the coil shown in figure 5.8, it has the effect of magnetising a

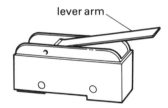

Figure 5.6 One type of microswitch

Figure 5.7 A rotary switch

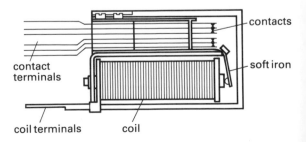

Figure 5.8 An electromagnetic relay

soft-iron plate which is drawn to the coil and opens and closes the contacts. These contacts can be used to switch a larger current than the current which flows through the coil. For example, it is necessary to use a relay if you want your computer to control motors and lamps.

(f) Reed switch This type of switch has two flat, flexible ferromagnetic reeds which are easily magnetised and demagnetised — see figure 5.9. The contacts are protected by being sealed inside a glass envelope which contains an inert gas such as nitrogen to reduce corrosion of the contacts. If a permanent magnet is brought close to the reeds, the reeds are magnetised, attract each other and 'make' an electrical circuit in series with the reed switch. On removing the magnet, the reeds lose their magnetism and therefore separate, and the circuit is opened.

The reed switch has numerous applications, since it has the following advantages over other forms of switch.

(i)   It is a *proximity switch*, meaning that it is the nearness of the magnet which alone causes the reeds to close. This 'action at a distance' enables the switch to be operated through a shield, provided that the shield is not a material which can be magnetised.

(ii) Since the contacts are sealed in an envelope, reed switches are ideal for use in atmospheres containing explosive gases. The sparking which occurs at the contacts of an unsealed switch could lead to an explosion when such gases are about.

(iii) It is a fast switch; some types of reed switch can be operated 2000 times per minute.

(iv) It has a long life. Up to 1000 million switching operations can be obtained from a reed switch, provided it is switching a low current. High currents rapidly cause deterioration of the contacts.

5.3   Uses for a reed switch

(a) Position control The diagram in figure 5.10 shows the principle employed here. The object being positioned carries a magnet, and the reed switch is fixed at the position at which the object is to stop. When the magnet is near the reed switch, the reeds close, and this is arranged to switch off the power to the positioning motor. This application could be used in controlling a model train.

(b) Measuring rotational speed Figure 5.11 (overleaf) shows how a rotating shaft, which is carrying a magnet, operates a fixed reed switch. The reed switch is a fast switch, so the shaft can rotate at up to 2000 rev/min and each rotation of the shaft will cause the reeds to close. The pulses from the reed switch can be used to give a rev/min reading on the dial of a meter. The resulting instrument is called a tachometer.

(c) Rotary-position indicator Figure 5.12 (overleaf) shows the principle. A magnet on

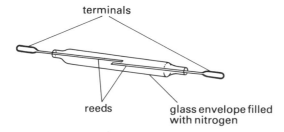

Figure 5.9 A reed switch

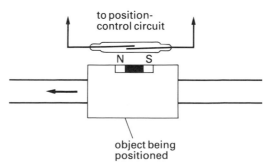

Figure 5.10 A reed switch used for position control

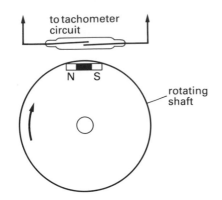

Figure 5.11 A reed switch used for measurement of speed

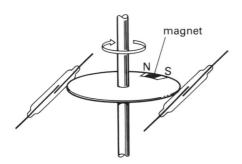

Figure 5.12 How a rotary position indicator works

a rotating shaft closes the reeds of the nearest reed switch. This closed reed switch can be used to operate a lamp or sound an alarm, so that the position of the shaft can be found without actually seeing it. A weather-cock could be designed making use of this idea. Four reed switches would give the four main points of the compass, but normally you would also want to record points like east-north-east. How many reed switches would you need for recording compass points that include east-north-east and south-south-east?

5·4 **Simple digital circuits**

Figure 5.13(a) shows a simple on-off circuit. If the switch if CLOSED the lamp is ON. If the switch is OPEN the lamp is OFF. This is a simple *digital circuit* since there are just two *states* for the switch (CLOSED or OPEN) corresponding to the lamp being

ON or OFF respectively. Figure 5.13(b) is the black box diagram of the circuit. The two possible states of the switch and of the lamp are given the *binary numbers* 0 and 1. Binary means 'two' and in the binary number system there are just two digits, 0 and 1, from which binary numbers are made up. Here we have chosen the digit 1 for the states 'lamp ON' and 'switch CLOSED', and the digit 0 for the states 'lamp OFF' and 'switch OPEN'.

The table in figure 5.13(c) is called a *truth table* since it shows all possible relationships between the input and output information. Truth tables are a useful way of looking at the function of digital circuits and systems.

Now let's connect a second switch B in series with the first switch A as shown in figure 5.14(a). Note that this is a simple series circuit since the switches and the

(a)

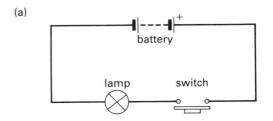

(b)

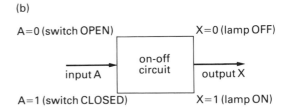

(c)

input A	output X
switch OPEN A=0	lamp OFF X=0
switch CLOSED A=1	lamp ON X=1

Figure 5.13 (a) A simple on-off circuit. (b) The circuit is shown as a functional black box. (c) The truth table for this on-off circuit

lamp are connected together one after the other. The lamp cannot light up unless switch A *and* switch B are closed. As the black box shows (figure 5.14b), the switches provide two binary digits of input information and the lamp indicates the value of the output information. The truth table summarises the function of this series circuit which is known as an *AND gate* since the output state has a value of binary 1 (lamp on) only if switch A *and* switch B each have a value of binary 1 (both switches closed).

Note that these digital circuits are called *gates* because they open and close to control the information reaching the output. The gate is open if the output is binary 1, and closed if it is binary 0.

Another simple digital circuit is shown in figure 5.15(a). You will recognise that the switches, A and B, are connected in parallel. In this circuit the lamp will light if switch A *or* switch B is closed. The lamp will also light if both switches are closed. This *OR gate* has the truth table shown in figure 5.15(c) and gives the value of the output information for all possible values of the input information. Strictly this gate is known as an *inclusive OR gate* since it includes the condition that the output has a binary value of 1 and not 0 when both input values are 1. An *exclusive OR gate* excludes this last condition so the last row in figure 5.15(c) would give an output value of 0.

The branch of electronics introduced above is known as *digital logic*. AND and OR

(a)

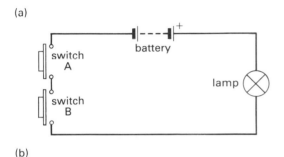

(b)

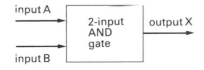

(c)

input A	input B	output X
0	0	0
1	0	0
0	1	0
1	1	1

Figure 5.14 (a) A simple series circuit. (b) The circuit shown as a black box which has two inputs and one output. (c) The truth table for this 2input AND gate

(a)

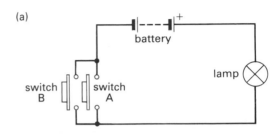

(b)

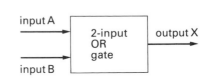

(c)

input A	input B	output X
0	0	0
1	0	1
0	1	1
1	1	1

Figure 5.15 (a) A simple parallel circuit. (b) The circuit shown as a black box (which has two inputs and one output). (c) The truth table for this 2input OR gate

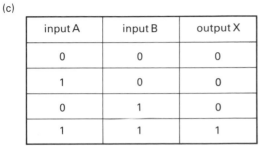

gates are called *logic gates* since the output state is a logical (i.e. predictable) result of a certain combination of input states. Logic gates are of great importance nowadays in the design of digital circuits and systems. You will find a more detailed description of them in Chapter 3 in Book E.

Questions

1   Figure 5.16 shows a circuit with three switches in it. Produce a truth table for this circuit which shows the state of the lamp (1 for ON and 0 for OFF) for all possible combinations of the input states set by the switches (1 for CLOSED and 0 for OPEN).

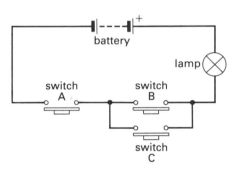

Figure 5.16 A series-parallel digital circuit

2   Figure 5.17 shows how two single-pole, double-throw switches are used to enable a light to be switched on from two positions. Explain how the circuit works.

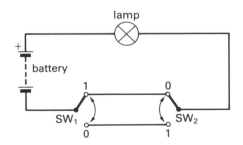

Figure 5.17 A two-way switching circuit using two single-pole double-throw switches, SW_1 and SW_2

3   The two positions for each of the switches in figure 5.17 are marked with the binary values 0 and 1. Draw up a truth table for this circuit. For lamp ON, use the binary number 1: for lamp OFF, use the binary number 0. What type of logic gate does this truth table represent?

4   Figure 5.18 shows how a double-pole, double-throw switch, SW_1 is used to change the direction of current flow through a d.c. motor, M. Trace the path of the current in the circuit for each position of the switch and show that the current though the motor reverses direction when the switch is operated. Note that the dotted line means that the two switch arms move together.

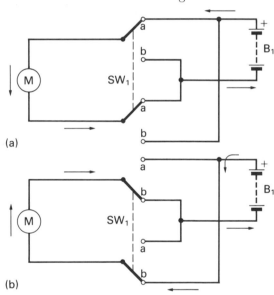

Figure 5.18 A reversing circuit for an electric motor using a double-pole double-throw switch

5.5 *Experiment* A4

Designing a two-way light switch circuit

Use a length of terminal block as shown in figure 5.19 to assemble a two-way light-switch circuit. Instead of using two single-pole, double-throw switches the wires

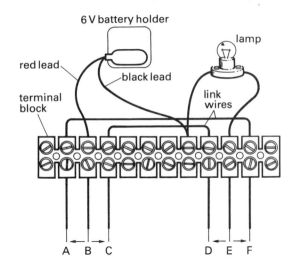

Figure 5.19 A two-way switching circuit using terminal block

A to F are connected together as follows: Connect together the bare ends of wires A and B and also the bare ends of wires E and F. The lamp will light. Now switch off the lamp by connecting together wires B and C. The lamp can be lit again by connecting wires E and D. Experiment with the connections of these wires until you understand how this two-way switching circuit works.

CAUTION

You should not wire up a mains-operated two-way circuit, say for a stairway or passage way, unless you get help from someone who is qualified to make connections to the mains supply.

5.6 *Experiment* A5

Designing a d.c. motor reversing circuit

Use a length of terminal block as shown in figure 5.20 to assemble a circuit for reversing the direction of current through a d.c. motor. If you connect the bare ends of wires B and C together and the bare ends of wires B' and C', the motor will rotate one way. Then, if you connect the wires B and A together and the wires B' and A' together, the motor will rotate in the opposite direction. As shown in figure 5.18, a double-pole double-throw switch can be used to reverse the direction of rotation of the motor.

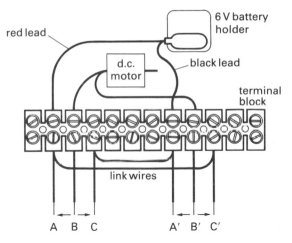

Figure 5.20 A reversing circuit using terminal block

6 Putting Circuits Together

6.1 The breadboard method

The *terminal block* used in Section 4.4 is an example of a solderless method for connecting components together. It has the advantage that the circuit can be dismantled quickly and the components used again for another circuit.

The terminal block is suitable only for simple circuits which do not use integrated circuits (ICs). Since many of the circuits in this book are based on ICs, a solderless *breadboard* of the type shown in figure 6.1 is generally used. This board has holes spaced one tenth of an inch (2.54 mm) apart which is the distance between adjacent pins on an IC. On the breadboard shown, several ICs can be inserted into the holes so that each IC straddles the central channel. Access to the pin connections on the IC is made by inserting the leads of components into the holes which are at right angles to the central channel. These holes are grouped in fives, each group being connected together underneath the holes

by a simple spring clip made of nickel-silver alloy.

Any component lead which is pushed into a hole is gripped by one of the five inter-connected spring clips thus making electrical contact with any other components which share the same spring clip. Power supply holes run the full length of both sides of the breadboard and access to these is made by jump leads from the groups of five holes. Jump leads are also used when making connections between different groups of five holes. To increase the working area, individual breadboards can be slotted together. An upright panel pushed into slots along one edge enables switches and potentiometers to be fitted.

The following notes will help you to use a breadboard:

(a) Use PVC covered 0.6 mm diameter wire for all wire links. The PVC should be stripped from the ends of the wire for about 8 mm.

(b) Make component leads and jump leads as short as possible.

(c) Make sure that component leads are free of kinks and blobs of solder otherwise the leads may jam in a hole and damage the spring clip.

(d) As far as possible arrange the components so that the input and output sections of the circuit are at opposite ends of the board.

(e) Double check your circuit layout against the circuit diagram before connecting the power supply to the circuit.

(f) The workpad or designer's sheet available with most breadboards ought to be used for planning the layout of the circuit before assembly.

Figure 6.1 A solderless breadboard assembly system — 'Verobloc'
Courtesy: Bicc-Vero Packaging Ltd.

6.2 **Soldered circuits**

Circuits assembled on breadboard can be modified and tested until the circuit works to your satisfaction. But the breadboard assembly is not intended for a permanent project and a much more rugged and less expensive assembly can be obtained by soldering the components together. The three main methods for making soldered circuits are *matrix board, stripboard* and *printed circuit board (PCB)*.

Matrix board is a simple system in which components are soldered between pins pushed through holes in an insulating board called *synthetic resin-bonded paper (SRBP)*. There is freedom to position the pins where you want to achieve a circuit layout to your taste and that is best suited to component sizes. Its great advantage compared with stripboard (see below) is that all the circuit connections can be seen from one side of the board. The holes are separated by 0.1 inch (2.54 mm) to allow integrated circuits to be used.

Stripboard has a number of parallel copper tracks bonded to SRBP. The copper tracks are pierced with holes for the component wires. The separation of these

holes along the copper track and between adjacent tracks is 0.1 inch, precisely. Like breadboard, this 0.1 matrix stripboard allows integrated circuits to be used whose pins are spaced 0.1 inch (2.54 mm) apart.

Figure 6.3 (overleaf) shows how stripboard is used. The components are positioned on the board by poking their leads through the holes from the side opposite to the copper tracks. It is a good idea to position all the components before soldering. The components and link wires can be prevented from falling out as you turn the board over by bending their leads slightly on the copper track side of the board. A small bench vice is quite useful for holding the board during this stage of the operation and for the soldering stage that follows.

Before soldering, you should consider the merit of cutting the copper tracks at the points where sections of the same track have to be isolated. If you do not cut the track until after soldering the wires in place, you may find solder has run along the track and blocked the hole where the track is to be cut.

To make a good soldered joint, the tip of the soldering iron must be in contact with both the copper track and the wire to be soldered to the track. The solder should flow off the tip of the soldering iron and around the wire to be soldered without going as far as the two adjacent holes. This operation should take no more than about five seconds otherwise the heat conducted through the wire might damage the component or weaken the bond between the copper and the board. It is not necessary to use a *heatsink* to prevent a component, such as a transistor, from getting too much heat. Most semiconductor devices nowadays can tolerate getting hot and, in addition, integrated circuits are plugged into a holder after the holder has been soldered in position.

Figure 6.3(b) shows the correct appearance of a soldered joint; it should appear bright when cool and the solder should have 'wetted' the copper track, stopping just short of neighbouring holes.

Figure 6.2 Using a matrix board

Sidecutters should then be used to cut off the waste wire close to the soldered joint.

Finally, after all the components and wire links have been soldered in place,

a)
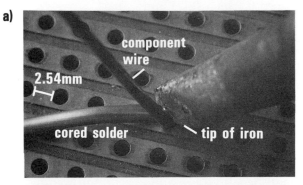

b)

c)

d)
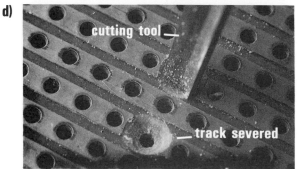

Figure 6.3 Stages in the making of a soldered joint

check over all the joints using an eyeglass or magnifying glass — don't be in too much of a hurry to connect the power supply to the circuit and see if it works! You should look for three possible problems: wires you have omitted to solder to the track; tracks not severed completely; and solder which has run between adjacent tracks. To remedy the latter problem, wipe excess solder off the tip of the iron and move it smartly between the tracks melting the solder on the way and this should separate the solder. If not, clean the iron and repeat the action.

6.3 Printed circuit board assembly

Background

Circuits in which components are connected together using copper tracks on an electrically insulating board are known as printed circuits. The *printed circuit board (PCB)* was invented during the World War 2 by Paul Eisler in an attempt to reduce complicated wiring in electronic equipment. PCBs were not very successful in valve circuits — see Chapter 2 — but when transistors came along in the 1950s the PCB became worthwhile on a commercial scale and is now commonplace for wiring circuits which include integrated circuit packages.

The basic material used in the making of PCBS is a copper-clad laminate. This material consists of a thin film of copper bonded to a base material (the 'board'), which is usually epoxide-impregnated glass fibre. A PCB is made by etching away the copper film to leave a predetermined pattern of copper tracks on the laminate. This pattern is produced by printing the pattern of circuit connections on the laminate using a material which is *etch resistant*.

Given a circuit, it is possible to design the artwork for making a PCB. However, the procedure described below shows how to make the PCB from artwork already provided, as it is for each of the Project Modules described at the end of the five books in *Basic Electronics*.

The PCBs of the Project Modules are produced by a *photographic process* using *ultraviolet light*. A suitable ultraviolet (UV) unit is shown in figure 6.4. UV units are generally expensive so it is a good idea to get the loan of one from a local technical college or polytechnic, especially an institution which has an art department since UV units are used to produce silk screens for printing. The UV light which is produced in these units is filtered long-wave radiation and not the shortwave variety used to erase EPROMs (see Section 15.6, Book E).

CAUTION

Though the UV light from these exposure units is not dangerous, avoid looking directly at the tubes because your eyes may begin to ache after a while.

Materials

Once a suitable UV unit is available, the following materials are required to produce a printed circuit board from the artwork provided for the Project Modules in *Basic Electronics* — see Chapter 12.

Figure 6.4 A UV unit

(a) Fibreglass board This board should have a 35 micron (35 μm or 35×10^{-6} m) thick cladding of copper with a positive photoresist coating. *Positive photoresist* means that opaque lines on the master artwork produce identical copper tracks (after etching) on the copper clad board. Negative photoresist produces an opposite image. Note that a coating of positive photoresist is protected from the light by a black plastic film which is removed before exposure to UV light.

(b) Developers
(i) For the photoresist, use sodium hydroxide crystals. One 80 g sachet of crystals dissolved in about 4 litres of water makes up a 2% solution suitable for developing the image on the PCB after exposure to UV light.
(ii) For the exposed copper laminate, use ferric chloride crystals. One 2 kg pack of crystals makes up 4 litres of saturated solution (known as the *etchant*) and is suitable for removing exposed copper from positive photoresist boards.

Copying the artwork

Before using the UV unit, a copy of the original artwork, e.g. the PCB pattern of the Pulser described in Section 12.5, must be made on tracing paper. First pencil in the outline of the artwork and then produce a black ink copy of it on the tracing paper. This copy is sometimes called the PCB mask. Make sure that the mask is a perfect copy of the original artwork and that there are no broken or joined tracks on it.

Preparing the copper laminate

All the Project Modules in *Basic Electronics* are prepared on 90 mm × 50 mm pieces of PCB. Once the PCB has been cut, any roughness on its edges should be removed with a file. DO NOT REMOVE THE BLACK FILM ON THE PCB YET.

Developing and etching the board

The following steps summarize the process of preparing the PCB so that components can be soldered to it.

(a) Switch on the UV unit for a few minutes to allow the tubes to warm up, and then switch it off. Do not allow the PCB to be exposed to bright sunlight or fluorescent lights. Preferably work in subdued light or yellow safelight.

(b) Place the PCB mask, i.e. the artwork you have prepared, on the glass of the UV unit. The correct way up for the mask is when any words on it are back to front.

(c) Peel off the protective black plastic film from the PCB, and carefully position the PCB over the mask with the photosensitive surface touching the mask. Carefully close the lid of the UV unit to make sure that the PCB and mask are not disturbed.

(d) Switch on the UV unit and expose the PCB for the recommended time for the unit. Typical exposure times are between two and six minutes.

(e) Switch off the unit, and place the exposed PCB in a tray of the developer, i.e. the sodium hydroxide solution. Use a soft brush or cotton wool to remove the developed photoresist. It helps if the developer is warmed slightly. You should see a colour change during development, the unexposed tracks showing up as greenish-blue. Deliberately overdevelop the PCB.

(f) Remove the PCB from the developer and carefully rinse it in cold water to remove all the excess photoresist coating. Leave the PCB to dry.

(g) Warm the ferric chloride solution by placing its container in a sink of warm water at about 45°C. It would be very difficult to remove copper from the exposed copper laminate if cold etchant was used.

(h) Place the developed PCB in the tray and pour in enough etchant to cover it. Constantly agitate the etchant to remove the exposed copper. This process takes about 20 minutes, or more if the etchant has been used before. It helps to speed up the process if the tray is placed in a sink of warm water. Keep a watch on the board to make sure that the etchant does not undercut the tracks.

(i) When the board has been completely etched, remove it and rinse it in cold water. The etchant may be returned to the container unless it has become weak, i.e. if etching has taken a long time.

(j) Inspect the PCB with a magnifying glass. Remove any bridges between tracks with a *reamer*. Broken tracks can be joined by running solder over the gap. Note that the undeveloped photoresist which covers the tracks need not be removed before soldering.

Drilling the PCB

Components are inserted through holes in the PCB and soldered to the copper tracks and pads. 1 mm diameter holes are needed for component leads; larger holes are required for bolts, e.g. for attaching terminal blocks, coils, transformers, and the like to the board. To avoid breaking small drills, use the more expensive tungsten carbide drills rather than high-speed steel drills. If possible, drill the holes using a drill stand to avoid damaging the drills. Make sure they are sharp and do not apply too much pressure to the lever on the drill otherwise a small burr round the hole will almost certainly produce a poor soldered-joint. Drill out all the holes with the 1 mm drills, and then use the bigger drill on those which need to be larger.

PCB assembly

When soldering the circuit on the drilled PCB, you should start with the smaller components first, such as resistors, polyester capacitors and wire links. Continue with the larger components. Make sure polarized components, e.g. Zener diodes, electrolytic capacitors and light-emitting diodes, are inserted in the board the correct way round. Do not apply too much heat when soldering. Repeated soldering and unsoldering will damage components and lift the copper track.

6.4 **Tools for projects**

Newcomers to electronics are often bewildered by the great many tools shown in component distributer's catalogues. In fact, to assemble circuits and 'house' them requires only a few handtools and these are listed in figure 6.5 as 'essential' and 'non-essential'. You will find, too, that you don't need a large area in which to work and store components and materials.

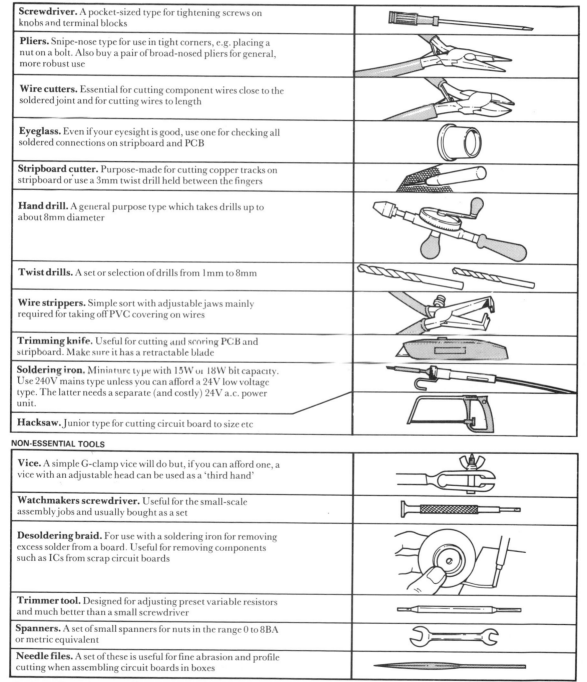

ESSENTIAL TOOLS

Screwdriver. A pocket-sized type for tightening screws on knobs and terminal blocks	
Pliers. Snipe-nose type for use in tight corners, e.g. placing a nut on a bolt. Also buy a pair of broad-nosed pliers for general, more robust use	
Wire cutters. Essential for cutting component wires close to the soldered joint and for cutting wires to length	
Eyeglass. Even if your eyesight is good, use one for checking all soldered connections on stripboard and PCB	
Stripboard cutter. Purpose-made for cutting copper tracks on stripboard or use a 3mm twist drill held between the fingers	
Hand drill. A general purpose type which takes drills up to about 8mm diameter	
Twist drills. A set or selection of drills from 1mm to 8mm	
Wire strippers. Simple sort with adjustable jaws mainly required for taking off PVC covering on wires	
Trimming knife. Useful for cutting and scoring PCB and stripboard. Make sure it has a retractable blade	
Soldering iron. Miniature type with 15W or 18W bit capacity. Use 240V mains type unless you can afford a 24V low voltage type. The latter needs a separate (and costly) 24V a.c. power unit.	
Hacksaw. Junior type for cutting circuit board to size etc	

NON-ESSENTIAL TOOLS

Vice. A simple G-clamp vice will do but, if you can afford one, a vice with an adjustable head can be used as a 'third hand'	
Watchmakers screwdriver. Useful for the small-scale assembly jobs and usually bought as a set	
Desoldering braid. For use with a soldering iron for removing excess solder from a board. Useful for removing components such as ICs from scrap circuit boards	
Trimmer tool. Designed for adjusting preset variable resistors and much better than a small screwdriver	
Spanners. A set of small spanners for nuts in the range 0 to 8BA or metric equivalent	
Needle files. A set of these is useful for fine abrasion and profile cutting when assembling circuit boards in boxes	

Figure 6.5 Basic tools for circuit assembly

7 Atoms and Ions

7.1 The building blocks of atoms

Nearly all materials on Earth are in solid, liquid, or gaseous form and whatever form these materials take, they all consist of atoms. *Electrons* are an important part of an atom, but so are *neutrons* and *protons*. These are the three main particles from which atoms are made.

Atoms are extremely small 'bits' of material — millions of them lie side-by-side across the diameter of the ink dot at the end of this sentence. Nevertheless, they can be seen, or at least their general position can be made out, as figure 7.1 shows.

Although atoms are so small, a great deal is known about them. For instance, how they combine with one another to produce *molecules*, and what makes some atoms *stable* and some *unstable*.

Figure 7.1 Field-ion microscope picture showing positions of atoms
Courtesy: Science Photo Library/Dr. Erwin Müller

It is not an easy matter to break an atom into pieces, or even to build one up out of protons, neutrons and electrons. However, the fact that *radioactive* materials exist in and around us proves that some atoms are naturally unstable and are breaking up of their own accord.

Uranium is a radioactive material and atoms of it are all the time breaking up and changing into other types of atom. When a large number of uranium atoms are deliberately encouraged to break up, as in the nuclear reactor of a nuclear power station, a great deal of heat is produced which is used to generate electricity.

All atoms have a *nucleus*. It is the nucleus which changes when an atom is radioactive. The proton and the neutron have their home in the nucleus, but the electron is to be found outside the nucleus, making up what is called an *electron cloud*. The nucleus is very small compared with the general size of an atom — say the size of an orange compared with the vast volume of an English cathedral. Using this model for an atom, you can imagine the electrons to be flies in the cathedral. Most of the mass of an atom lies in the nucleus. In fact, the proton and the neutron have about equal masses, but the electron has a mass about 2000 times smaller than either of these particles.

7.2 The electrical charge on neutrons, protons and electrons

The most important property of these three particles, from the point of view of our study of electronics, is their *electrical charge*. The electron carries a negative charge and the proton a positive charge. These charges are equal in size but electrically opposite. Since the charges are opposite, electrons and protons attract each other. It is this attraction which keeps electrons in a cloud around the nucleus although they don't actually fall into it. Each electron in an atom can possess only a certain value of energy, and the electrons arrange themselves into *shells* according to the energy they possess. Thus, each shell contains electrons of a particular energy.

The neutron does not carry any electrical charge, i.e. it is neutral. Although neutrons do not have any part to play in keeping the electrons in their shells, they are an important part of an atom since they contribute to the mass of an atom. But it is the number of protons which determines the type of material to which the atom belongs, not the number of neutrons.

7·3   The building blocks of hydrogen and oxygen atoms

Hydrogen and oxygen are two very common elements, since their atoms go to make up that very useful and vital liquid called water. Actually, a hydrogen atom has the simplest structure of all. Figure 7.2 shows two kinds of hydrogen atom to be found in nature, known as hydrogen-1 and hydrogen-2. These two types are called *isotopes* of hydrogen. Water made from

hydrogen-2 (also called *deuterium*) is chemically the same as ordinary water, and you can drink it, grow plants in it and swim in it. But it is slightly heavier than ordinary water, and it is called *heavy water*. Heavy water also boils at a little higher than 100°C and freezes at a little higher than 0°C. Natural water contains a very small quantity of this heavy water. Note that both these kinds of hydrogen have an equal number of protons (one) in the nucleus, and it is this number which tells us that the atom is the element hydrogen.

The atom of oxygen is more complex than that of hydrogen. It has eight protons in its nucleus, but it can have seven, eight, or nine neutrons within it without changing the chemical properties of the atom. Figure 7.3 shows the structure of the most common type of oxygen atom, having eight protons and eight neutrons in the nucleus. Eight electrons make up the normal oxygen atom, and they are arranged in two shells.

Atoms which have equal numbers of protons in their nuclei but different numbers of neutrons are known as isotopes of that element. You may have heard of radioactive isotopes in connection with their use in medicine, agriculture and industry. One of the most important points for you to note is that in an atom in a normal state, that is, a neutral state, the number of electrons in the shells around the nucleus is equal to the number of protons in the nucleus.

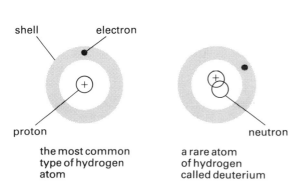

Figure 7.2 Two kinds of hydrogen atom

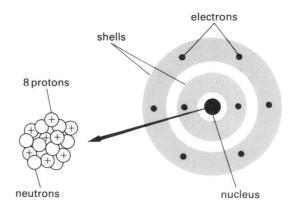

Figure 7.3 Structure of an oxygen atom

Questions

1 The table shown in figure 7.4 lists a few kinds of atom, two of which, silicon and germanium, are discussed in Sections 7.5 and 7.6. Can you fill in the blank spaces?

7.4 **Free electrons**

Protons and neutrons are bound together very tightly in the nucleus. Despite the fact that neutrons have no electrical charge, there is a very powerful close-range force which acts between a neutron and a proton. The strength of this *nuclear force* accounts for the large quantities of energy generated when the nuclei of uranium atoms split apart in an atomic fission bomb or a nuclear reactor.

However, the force between an electron in an atom and the protons in the nucleus is much weaker than the nuclear forces — it is upon this weakness that electronics is based. An electrical current is caused by the movement of electrons. Some of the electrons of the atoms of silver and copper (two electrical conductors) are so weakly bound, that these electrons are called *free electrons*. However for other materials (the

Figure 7.4 Some information about atoms

atom	no. of protons in the nucleus	no. of electrons in the shells	no. of neutrons in the nucleus	no. of protons and neutrons in the nucleus
Hydrogen–1	1	1	0	1
Hydrogen–2	1	1	1	2
Oxygen–16	8	8	8	16
Oxygen–18	8	—	10	18
Copper–63	29	29	—	63
Silver–108	—	47	—	108
Silicon–28	—	—	14	28
Germanium–74	32	—	—	—
Carbon–12	—	6	—	—
Carbon–14	—	—	—	—
Iron–56	26	—	—	—

electrical insulators) the electrons are less easily pushed around; they are more strongly bound to the atoms. Atoms which have more or less electrons than normal are called *ionised atoms*.

Although an electron will gladly settle down into one of the shells of an atom if there is place for it, in the emptiness of Space, electrons are found moving vast distances without coming near enough to nuclei to form a neutral atom. Nearer home, we find electrons existing for a short time separate from a nucleus in a television picture tube. The picture on the screen is 'written' by a fast-moving stream of electrons, as is the trace which is seen on the screen of an oscilloscope.

Questions

1 Why is there very little air in a television or oscilloscope tube?
2 What other devices do you know which make use of free electrons moving in a vacuum?
3 What do you think happens to the electrons as soon as they strike the screen of a television tube?

▽ 7.5 **Conductors, insulators and semiconductors**

The reason why some materials, such as copper, are good electrical conductors is that they contain 'free' electrons which are quite weakly bound to the nuclei of the atoms of the material. These electrons can be moved easily by connecting the material across a battery. Copper and aluminium are good electrical conductors and are used in electronics to allow electrons to flow easily between one device and another. Electrons are more strongly attracted to their parent nuclei in electrical insulators, which therefore do not have any free electrons. Thus electrical insulators such as glass, polythene and mica are used to resist

7.6 **Silicon atoms**

the flow of electrons between electronic devices.

Electronics is to do with the use of semiconductors as well as conductors and insulators. Semiconductors are the basis of electronics devices such as transistors and diodes, heat sensors and light emitters, integrated circuits and many other devices. As its name suggests, a semiconductor has an electrical resistance that falls somewhere between that of a conductor and that of an insulator.

Two of the commonest semiconductors are silicon and germanium. They are important in electronics because their resistance can be controlled to good effect. There are two ways of doing this. First there is the effect of heat on a semiconductor. At very low temperatures semiconductors happen to be good electrical insulators. But as their temperature increases they become increasingly better electrical conductors so that at everyday temperatures they allow some current to flow through them. Generally, this drop in resistance with temperature increase is a nuisance, although some devices, such as thermistors (Book C), do make use of this effect.

The second way of controlling the electrical resistance of silicon and germanium is to add minute amounts of carefully selected substances to them. We need to know something about the atomic structure of germanium and silicon to understand the effect this has.

Germanium is now rarely used in electronic devices, which are mostly based on silicon. Figure 7.5 shows a model of a silicon atom that has fourteen electrons surrounding a nucleus containing fourteen protons and fourteen neutrons. The part of this structure that makes silicon useful to electronics is the way the electrons are arranged in what are known as shells surrounding the nucleus. There are two electrons in the inner shell, eight in the next shell, and four in the outer shell. It is the four electrons in the outer shell, known as the valency shell, which make pure silicon a crystalline material.

In a crystal of pure silicon, each of the four outer electrons forms what is known as a covalent bond with an electron from a neighbouring silicon atom. Figure 7.6, shows how the pairing of electrons uses up every one of these outer electrons. An orderly arrangement of silicon atoms results and gives pure silicon its crystalline structure. There are no free electrons available to make pure silicon conduct electricity and so it is an insulator. At least, it is an insulator at low temperatures, and a perfect insulator at the absolute zero of temperature ($-273°C$). But at everyday temperatures, silicon conducts electricity a little; not much but enough to make silicon a bit of a problem when it is used in transistors. However, we are not so much interested in how an increase of

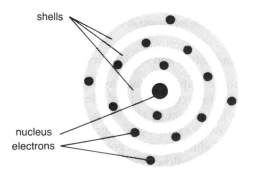

Figure 7.5 A silicon atom

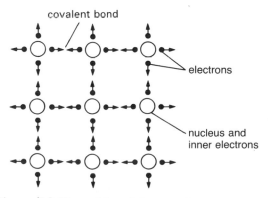

Figure 7.6 The pairing of electrons in atoms

temperature reduces the resistance of silicon, but in what happens to its resistance when a small amount of an △ 'impurity' is added to it.

▽ 7.7   n-type and p-type semiconductors

Once a very pure crystal of silicon has been manufactured, the silicon is 'doped' with impurity atoms! These atoms are chosen so that they make a 'bad fit' in the crystal structure of silicon, due to the impurity atoms having too many or too few electrons in their outer shells. Depending on the impurity, two types of semiconductor are produced in this way, n-type or p-type.

An *n-type* semiconductor is produced by doping silicon with, for example, phosphorus. A phosphorus atom has five electrons in its outer shell. Figure 7.7 shows what happens when an atom of phosphorus is embedded in the crystal structure of pure silicon. Four of the five outer phosphorus electrons form covalent bonds with neighbouring silicon atoms, leaving a fifth unpaired electron. This unattached electron is now weakly bound to its parent phosphorus atom and it is therefore free to wander about. Phosphorus is said to be a *donor* impurity since each atom of phosphorus can donate (give away) an

electron. The addition of phosphorus has therefore changed the electrical properties of silicon. It has become an electrical conductor due to the presence of free electrons donated by phosphorus atoms. An n-type semiconductor has been produced ('n' for negative).

A *p-type* semiconductor is produced by doping silicon with atoms such as boron which have three electrons in their outer shells. Figure 7.8 shows what happens when a boron atom becomes embedded in the crystal structure of silicon. Three of its outer electrons become paired with neighbouring silicon atoms, leaving one unpaired silicon electron. This electron is not available for conduction but it will accept another electron to pair with it. The vacancy created in silicon by doping it with boron is known (not surprisingly!) as a 'hole'. Since this hole attracts an electron, it behaves as if it had a positive charge. Boron is said to be an *acceptor* impurity since its atoms can accept an electron from other nearby atoms. The presence of holes which act as positive charges in boron-doped silicon produces a p-type semiconductor ('p' for positive). You will see how the behaviour of electrons and holes in n-type and p-type semiconductors accounts for the way diodes and transistors △ work in Book C.

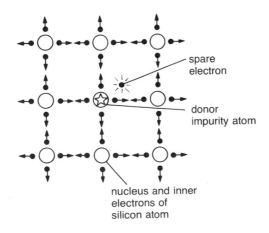

Figure 7.7 How a donor atom produces free electrons

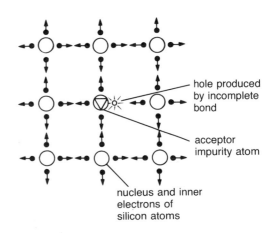

Figure 7.8 How an acceptor atom produces a hole

8 Measuring Signals

8.1 Three important measuring instruments

Successful designing and testing of electronic circuits depends on the intelligent use of one or more test instruments. This section describes three of these instruments — the *multimeter*, the *oscilloscope* and the *waveform generator* which are discussed in Chapters 9, 10 and 11.

A *multimeter* measures the strength of three very important electrical quantities: electrical current which is measured in *amperes*, electrical potential difference which is measured in *volts*, and electrical resistance which is measured in *ohms*. These three quantities — the ampere, the volt, and the ohm — were introduced in the first half of this book and they are discussed more fully in Book B.

An *oscilloscope* can also be used to measure the strength of an electrical quantity, but its main purpose is for showing the *shape* (the waveform) of rapidly changing electrical signals such as those produced by a microphone. These waveforms are displayed on the screen of the oscilloscope, just as computer graphics are displayed on the screen of a television. From this display it is possible to measure the strength (amplitude) and frequency of the waveforms.

The *waveform generator* (or *signal generator*) produces waveforms of different shapes and at different frequencies. These waveforms are used for testing the way circuits, e.g. audio amplifiers in hi-fi systems, respond to varying signals sent through them.

Without the use of these three instruments, particularly the multimeter which no circuit designer should ever be without, fault-finding becomes a matter of trial and error. If a circuit doesn't work as it is supposed to, you might start by replacing a component suspected of being faulty. You just might strike lucky and correct the fault in this way. But this procedure can be very time-consuming and components are easily damaged by being incorrectly connected in the circuit. The trial-and-error method is just not the way to learn electronics. This is where an instrument such as a multimeter is so helpful: it enables you to 'see' electrical signals which our senses are normally blind to. In addition measurements can be taken so that the performance of the circuit can be gauged and compared to see whether it meets the designer's specification.

8.2 The difference between d.c. and a.c. signals

The electrical signals which circuits produce can be broadly divided into *direct current (d.c.)* signals and *alternating current (a.c.)* signals. The multimeter and the oscilloscope are designed to measure the strength of both types of signal while the waveform generator is designed to produce only alternating current signals.

Direct current always flows in one direction in a circuit. The current produced by a torch battery is d.c. Current which keeps changing direction in a circuit is known as alternating current. The mains supply which comes into the home is a.c.

Figure 8.1 (overleaf) shows graphically how the steady direct current produced by a torch battery varies with time. As time passes, the current remains at a steady level (amplitude) and it flows in the same direction through a lamp. Of course, the current will eventually fall to zero in the final stages of the battery's life. But

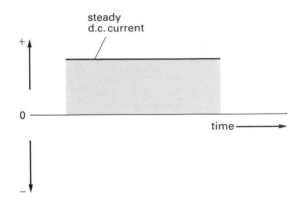

Figure 8.1 Graph of a steady direct current

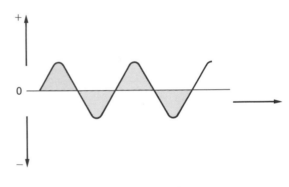

Figure 8.2 Graph of an alternating current

throughout its life, the battery produces current flowing in the same direction through the lamp it lights. Note that the plus (+) and minus (−) signs on the graph indicate the two possible directions of current flow through the circuit.

Figure 8.2 shows graphically how the alternating current from the mains supply flows first one way (in the positive direction), and then reverses (in the negative direction). Note that the strength (amplitude) of this a.c. waveform changes continuously as time passes and that the pattern repeats itself.

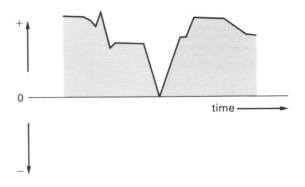

Figure 8.3 Current variation, a.c. or d.c.?

Questions

1   Is the current variation shown in figure 8.3, that of a d.c. change or an a.c. change?

2   Figure 8.4 shows a battery connected to two lamps, L_1 and L_2, by means of switches SW_1 and SW_2. Draw a graph to show how the current drawn from the battery varies when the switches are operated as follows: (a) both switches open for 10 seconds; (b) switch SW_1 closed for 5 seconds; (c) switch SW_2 closed for 20 seconds while SW_1 remains closed; (d) both switches open.

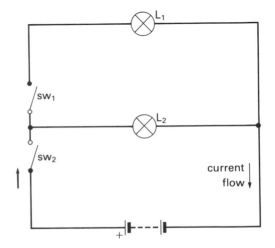

Figure 8.4 How does current vary as SW_1 and SW_2 are operated?

8.3 The frequency and period of alternating current

An alternating current is shown in figure 8.5. The current is first in one direction and then in the reverse direction, but notice that there is a smooth change of current from zero to a maximum in one direction and then to zero once again. It smoothly increases to a maximum in the opposite direction and then to zero again before repeating the cycle. This is the variation of current obtained from the mains supply and is known as a *sinusoidal variation* since the waveform has the shape of a sine wave. This variation of current is produced by a corresponding variation in the mains voltage which is called an a.c. voltage. The two terminals A and B (marked + and − in figure 8.5) show the variation of the a.c. voltage which occurs during each complete cycle of the waveform. For half the cycle the voltage on terminal A is more positive than on terminal B, and for the second half cycle terminal B is more positive than terminal A.

The time taken for the voltage at one terminal of the mains to change from its maximum positive value to its maximum negative value, and then back to its maximum positive value, is known as the *period* of the sine wave. For the a.c. mains voltage, the period T is 1/50 of a second:

$$T = 1/50 = 0.02 \text{ second}$$

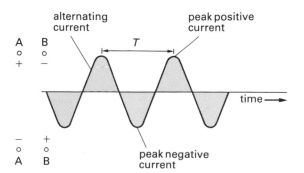

Figure 8.5 Graph of a sinusoidal alternating current

The number of complete periods of the sine wave in one second is known as the *frequency* (f) of the a.c. mains voltage, and is measured in *hertz* (Hz), which is given by

$$f = 1/T = 50 \text{ Hz}$$

The frequency of the mains supply very rarely differs from 50 Hz, and it is therefore used as a frequency standard. In fact, some equipment such as a mains-operated clock, depends for its accuracy on the steadiness of this frequency.

Questions

1   Is the current change shown in figure 8.6 that of a.c. or d.c.?

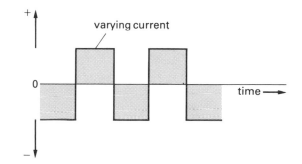

Figure 8.6 Is this a.c. or d.c?

2   What is the time occupied by one cycle of a radio wave of frequency 200 kHz?

The effect of the constantly reversing voltage, and hence current, is often noticed when using mains-operated equipment. You might notice a flicker from fluorescent tube lighting — the trick is not to look directly at the tube but to see it out of the corner of your eye. Sometimes a hum is heard from an oven when it is switched on, and occasionally a hum may be heard from a mains radio, expecially if the radio is not of particularly good quality.

▽ ## 8.4 Root-mean-square (r.m.s.) values of sinusoidal voltages

This description of an a.c. voltage or current often gives people difficulty. When we say that the voltage of the a.c. mains is 240 V, we mean that its *effective value* is 240 V. This figure is known as the r.m.s. voltage of the mains voltage and is the voltage which would give the same heating effect as a d.c. voltage of 240 V. However, the *maximum voltage* reached twice each period (see figure 8.5) is higher than 240 V. The root-mean-square voltage, V_{rms}, is related to the peak voltage, V_{max} by the equation

$$V_{rms} = 0.7 V_{max}$$

The relationship between V_{max} and V_{rms} is shown in figure 8.7. It is easy to show that the peak voltage of the mains is about 340 V. From the above equation,

$$V_{max} = V_{rms}/0.7 = 240/0.7 = 340 \text{ V}$$

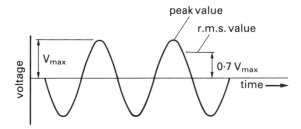

Figure 8.7 Meaning of peak and r.m.s. voltage

Questions

1 Calculate the r.m.s. value of a sinusoidally varying current which has a peak value of 3 amperes (3 A).

△

8.5 The importance of the earth connection

You know that the common method of connecting equipment to the mains is via a three-pin plug of the type shown in figure 8.8. Two pins are known as the *neutral* and *live* pins, and the third and thicker pin is the *earth* pin. The earth pin connects through the socket to the earth wire, which in turn is actually connected to the earth or ground outside a building. The symbol for an earth connection is shown in figure 8.9. The neutral wire is also connected to ground at the generating station, but actually a small voltage is often found to exist between the earth and the neutral pins. This small voltage is caused by currents which flow through the neutral wire from a number of items of equipment on the circuit. However, the live wire carries a high voltage (240 V r.m.s.) with respect to the neutral and the earth wires. It is across the earth and live wires that a voltage variation of the type shown in figure 8.5 is obtained.

The earth connection is used for safety reasons. The metal case of an oscilloscope, or a soldering iron, or a washing machine

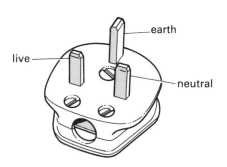

Figure 8.8 The pins of a mains plug

Figure 8.9 Earth symbol

could become 'live' if only the live and neutral connections are made, as shown in figure 8.10. This would mean that anybody touching the metal case and also having a good contact with ground would make a fairly good conducting path to ground. Sufficient current might flow through their body to kill them.

However, if the metal case if connected to ground — that is, earthed — and a fault occurs so that the live wire makes contact with the case, the fuse 'blows' because of the large current which flows through it. Thus the case cannot remain live, since the fuse blows first. The correct connections for the fuse and the earth are shown in figure 8.11.

When a new plug is fitted to a piece of equipment, always make sure that the live, neutral, and earth wires are connected correctly. There is a *colour code* for these three wires: brown for the live, blue for the neutral, and green and yellow stripe for the earth.

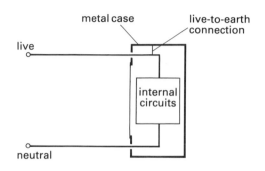

Figure 8.10 The danger of not having an earth connection

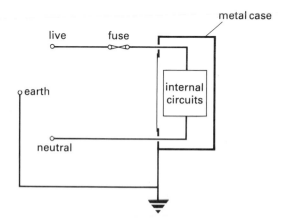

Figure 8.11 How an earth connection is made

9 The Multimeter

9.1 General description

This instrument is essential for both the amateur electronics enthusiast and the professional circuit designer. The multimeter combines in one portable, battery-operated unit the facilities for measuring the three electrical quantities volts, amperes and ohms. Both d.c. and a.c. voltage and current measurements can be measured over a wide range using a rotary switch, or push switches, to select the range and quantity required.

A multimeter is usually identified by the type of display it uses: it is made with either an *analogue* or a *digital* display. These two types of multimeter are shown in figure 9.1 and figure 9.2, respectively, and are referred to as an analogue multimeter or a digital multimeter.

A digital multimeter is easier and cheaper to make than an analogue multimeter, largely because its electronic circuits and its display are solid-state, i.e. designed using transistors, integrated circuits and light emitting diodes (LEDs) or liquid crystal displays (LCDs). LED and LCD displays are described in Book E. It does not have a display with moving parts like the analogue multimeter which uses a moving coil meter. The moving coil meter is a delicate mechanism and is made of a coil which is supported in bearings and rotates between the poles of a strong

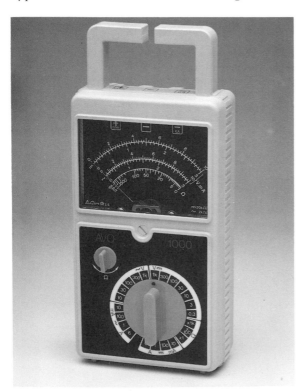

Figure 9.1 An analogue multimeter
Courtesy: Megger Instruments Ltd.

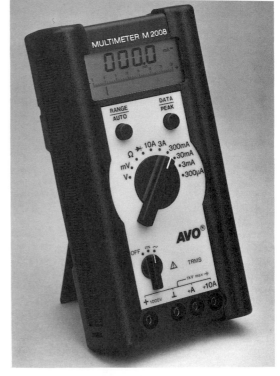

Figure 9.2 A digital multimeter
Courtesy: British Brown Boveri

magnet (figure 9.3). The pointer attached to the coil moves over a calibrated scale. The rotation of the pointer is restrained by spiral springs (sometimes the coil is supported by a thin wire as in a torsion band meter) which also carry the current to the coil (figure 9.4). Severe mechanical shocks can damage this delicate movement so don't ever drop an analogue multimeter.

The current that flows through the coil produces a magnetic field, as explained in Chapter 16 in Book B. This magnetic field makes the coil rotate in the much stronger magnetic field produced between the poles of the permanent magnet. The stronger the current through the coil, the stronger the

force acting on the coil and the greater the deflection of the pointer. The coil rotates until the force exerted by the springs equals the force produced by the magnetic effect of the current.

As with many other types of instruments (an electronic thermometer, for example), digital displays are becoming increasingly common for they are much less likely to be damaged by bumps and knocks when in use. In addition, a digital read-out gives more precise measurements of current, voltage and resistance. If the read-out makes use of an LCD (see figure 9.2) which requires negligible current to operate it, an internal battery will last for a very long time. A modern LCD multimeter also indicates the electrical measurement being made as a legend on the scale. Once a current range, say, has been selected, the user is reminded what quantity is being measured by the legend 'mA' or 'ohms' (figure 9.5).

Digital multimeters have many advantages compared with analogue multimeters but an analogue multimeter is useful in one particular way. Just as the use of hands on a watch is preferred since it is possible to relate time more easily to the past and future, an analogue multimeter is very useful for seeing changes in a measurement. The slow charging of a capacitor (Book B), for example, can be noted easily as the pointer of an analogue multimeter moves over a scale: it is much more difficult to see this trend when a lot of digits are changing on a digital display.

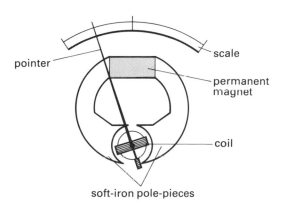

Figure 9.3 The main parts of a moving coil meter

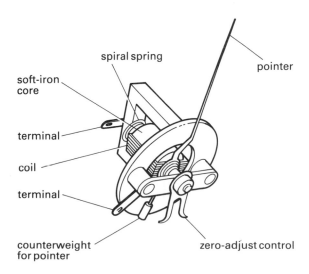

Figure 9.4 The positions of the spiral springs on a moving coil meter

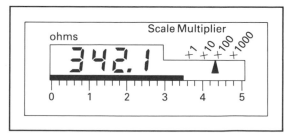

Figure 9.5 The legends available on one type of liquid crystal display

However, recent digital multimeters also have an analogue display built into them so that an increasing or decreasing trend in a measurement can be detected as well as the actual value of the measurement. As figure 9.5 shows, some LCD multimeters display a dark band which moves along a graduated scale and acts as an analogue read-out.

Today's digital multimeter does a lot more than simply measure the three basic electrical quantities. Both *frequency* and *temperature* can be measured easily. Whereas the maximum frequency that an analogue multimeter can measure is about 2 kHz, the digital multimeter can easily measure up to 100 kHz. Temperature is measured by plugging a unit coupled to a temperature sensor into the multimeter. Project Module C4 in Book C describes an add-on unit for temperature measurement. Some modern multimeters can act as *data loggers* and take measurements of a quantity (e.g. temperature) over a period of time and store these readings internally. Afterwards, these readings can be recalled on the display or fed into a microcomputer for display as a graph for further analysis. Figure 9.6 shows one of these sophisticated multimeters. Digital multimeters will continue to offer more functions and their cost is likely to fall, too. Analogue multimeters therefore look destined to become obsolete.

9.2 The place of ammeters and voltmeters in a circuit

An ammeter measures the flow of electrical current in amperes through a component. Ammeters must therefore be connected *in series* with the component. In figure 9.7, the ammeter is connected in series with the resistor R. Since the current has to flow through the ammeter as well as through the resistor, it must have as low a resistance as possible.

A voltmeter measures the potential difference, or electrical pressure, in volts across a component. Voltmeters must therefore be connected *in parallel* with the component. In figure 9.8, the voltmeter is connected in parallel with the resistor R. Since the presence of the voltmeter must not disturb the current flow through the resistor, it must have as high a resistance as possible.

Thus, ammeters must have a low resistance and voltmeters a high resistance

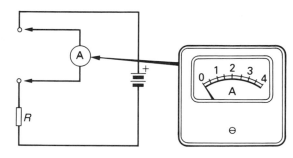

Figure 9.7 The position of an ammeter in a circuit

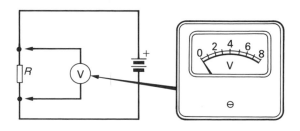

Figure 9.6 A modern multimeter and data logger

Figure 9.8 The position of a voltmeter in a circuit

if they are not to disturb the electrical quantities they are designed to measure. In fact, if you try to measure voltage across a component as in figure 9.8 with a multimeter switched to measure amperes, you are very likely to damage the meter. However, a good quality multimeter is protected by a fuse or by special electronic circuits, against this type of misuse. If you try to measure current through a component as in figure 9.7 with a multimeter switched to measure volts, no damage can result since the high resistance of the voltmeter reduces the current flow to a very small value.

9.3 The sensitivity of a voltmeter

Ideally, voltmeters should have an infinitely high resistance so that they do not draw any current from the circuit to which they are connected. A digital multimeter set to read ohms has an input resistance of at least 10 million ohms (10 MΩ) and it is constant on all ranges. Thus digital multimeters are almost ideal for measuring voltage.

However, the moving coil meter only operates if a current flows through its coil (see figure 9.4). This current is drawn from the circuit under test and alters the voltage which the multimeter is supposed to measure. If the meter has a resistance of 10

thousand ohms (10 kΩ), it cannot possibly give a proper value for the voltage across a 10 kΩ resistor in a circuit, since the meter seriously shunts the resistor. This multimeter is not very sensitive when measuring voltage.

The sensitivity of a moving coil voltmeter is measured in *ohms per volt:* the greater the ohms per volt rating the more sensitive the voltmeter. The ohms per volt (Ω/V) rating is given by

$$\frac{\Omega}{V} = \frac{1}{\text{current giving full-scale deflection (f.s.d.)}}$$

Thus, if a good quality moving coil meter requires 50 millionths of an ampere (50 μA) to give full scale deflection, its ohms per volt rating is 20 kΩ/V). In other words, when it is measuring a voltage of 1 volt, it draws a current of 50 μA. The sensitivity increases when higher voltages are measured since extra resistance is switched in series with the moving coil meter. Thus, on the 10 V range this sensitivity becomes 10 $\times$ 20 kΩ/V = 200 kΩ/V.

For general electronics work, do not choose an analogue multimeter with a sensitivity less than 20 kΩ/V. You may well see analogue multimeters which have sensitivities of around 200 kΩ/V. These instruments are expensive and rely on electronic circuits to increase their sensitivity, though the basic moving coil meter may only draw a current of 50 μA.

10 The Oscilloscope

10.1 General description

The oscilloscope is designed to show on its screen the shape of rapidly changing waveforms and to measure their frequency and amplitude. An oscilloscope usually operates from the alternating current mains supply, although an increasing number of portable, battery-operated oscilloscopes are being made for use in the field by engineers and others on the move. Most designers of electronic circuits think the oscilloscope is almost as indispensible a test instrument as a multimeter.

The proper name of a 'scope' is *cathode-ray oscilloscope* (CRO). The heart of a scope is a *cathode-ray tube*, similar to a television picture tube, inside which is an *electron gun*. This gun generates a thin beam (a ray) of rapidly moving electrons to 'write' a graph (a *trace*) on the screen of the tube of the waveform fed to its input. The trace might be of an alternating current, the signals produced by a microphone or amplifier, modulated radio waves, or the patterns of electrical signals produced by the heart.

A typical oscilloscope is shown in figure 10.1. This is a *dual trace scope* since the single beam of electrons generated by the electron gun is electronically switched at high speed to give the impression that there are two independent beams in the scope. The dual trace scope has two inputs so that two separate traces, derived from the single electron beam, can be 'written' on the screen. This enables waveforms to be compared. The switching of the beams must be much more rapid than the frequency of the waveforms at the inputs. The more expensive *dual beam oscilloscope* has two electron guns and it can produce traces of waveforms which have a much higher frequency.

Like a television picture tube, the inside surface of the screen of a scope is coated with a *phosphor*. This is a material which *fluoresces*, i.e. gives off light, when electrons strike it. Different phosphors give different colours of light, though green light is favoured for most scopes. The time for the light to fade away after the beam has passed is important too. For the general purpose scope used in electronics, the light must fade away quickly (in a few milliseconds) so that the trace does not linger and confuse the picture when rapidly changing waveforms are being observed. However, slowly changing waveforms, such as those generated by the heart, can only be seen clearly if the phosphor continues to emit light for a short time after the beam has passed.

Figure 10.1 A dual trace oscilloscope
Courtesy: Farnell International Instruments Ltd.

CAUTION

Make no attempt to probe the internal circuits of an oscilloscope unless you know what you are doing — some of the circuits carry lethally high voltages.

10.2 **The cathode-ray tube**

The cathode-ray tube gives a visual display of electrical waveforms, i.e. it converts electrical energy into light. It is used in cathode-ray oscilloscopes, television receivers, visual display units, radar, and other electronic systems. Figure 10.2 shows the basic structure of the type of cathode-ray tube used in an oscilloscope. This type of tube uses *electrostatic deflection*, rather than magnetic deflection, of the electron beam.

The electron gun contains a *cathode* made of a cylinder of nickel which is heated by a tungsten filament so that it gives off electrons — a process known as *thermionic emission*. The negatively charged electrons are accelerated towards the anodes A_1 and A_2 which are at a more positive voltage than the cathode. The strength of the electron beam reaching the anodes is controlled by the potentiometer, VR_1, which makes the *grid* more or less negative with respect to the cathode. Thus, VR_1 controls the brilliance of the trace formed on the screen.

The beam is focused on the screen to give a small luminous spot by varying the voltage between the anodes A_1 and A_2 using the potentiometer VR_2. A high voltage is used to accelerate and focus the beam of electrons generated by the electron gun. This voltage is shown in figure 10.2 as e.h.t.− and e.h.t.+. (e.h.t stands for *extra high tension*). For a small cathode-ray tube, the e.h.t. voltage on A_2 is about 1 000 V, on A_1 about 250 V, and on G between 0 V and 50 V.

To avoid a build up of negative charge on the screen, a conducting coating on the inside of the tube is earthed. On striking the screen, the electron beam produces secondary electrons which are attracted to the coating and return to the cathode via the external circuit. It is usual to earth the coating so that people and earthed objects near to the screen do not distort the direction of the electron beam. This means that the anode A_1 and the electron gun are negative with respect to A_2.

The movement of the spot of light across the screen is controlled by voltages applied across the X− and Y–*deflection plates*.

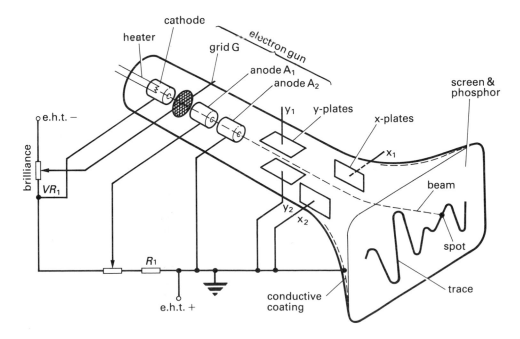

Figure 10.2 The basic structure of a cathode-ray tube in an oscilloscope

The voltage applied to the X-plates is generated by a *time-base amplifier*. Its job is to deflect the electron beam horizontally to make the spot sweep across the screen from left to right at a steady speed. This speed can be adjusted by the *time-base speed controls* on the scope. After each sweep, the time base amplifier switches off the beam and sends it back to the starting point at the left end of the screen — this process is known as *flyback*.

The waveform to be examined on the scope is amplified and applied to the Y-plates. The amplification of the waveform can be adjusted by *Y-sensitivity controls* on the scope. The input waveform causes the horizontal trace to be moved vertically in response to the strength of the waveform. A stable trace of the waveform only appears on the screen if each horizontal sweep of the trace starts at the same point on the left of the screen. This is achieved by feeding part of the input signal to a *trigger circuit*. This triggers (starts) the time-base when the input signal has reached a particular amplitude set by the *trigger level control*. Most scopes allow manual and automatic triggering of the time-base.

You may well find some differences in the layout of the controls on the front of the oscilloscope you use compared with the one shown in figure 10.1, but the controls are similar on whatever scope you are using. The following experiments will help you understand the function of these controls.

10.3 *Experiment* A6

Setting up an oscilloscope

(a)   Set the FOCUS and BRIGHTNESS (or BRILLIANCE) controls to about mid-scale, and switch the TIME/CM or TIME-BASE to the OFF position.
(b)   Set the INPUT (or Y-SENSITIVITY or VOLTS/CM) control to about 5 V/cm. Set the TRIGGER LEVEL to AUTO or INTERNAL.

(c)   If a control with a position marked TV is fitted, switch it to NORMAL.
(d)   Switch on the mains supply and allow the instrument to warm up. A green or blue blob should appear on the screen. If it does not, you will be able to find it by rotating the Y-SHIFT and X-SHIFT controls. Focus the spot. A control marked ASTIG will help, if one is fitted. Do not allow the spot to remain stationary for any length of time at full brightness or a 'burn mark' might be left on the screen.
(e)   The spot you have obtained is caused by a beam of electrons emitted from a hot filament in the CRO striking the fluorescent screen.
(f)   You are now able to obtain a sharply focused spot and to move the spot up and down and sideways by means of the X-and Y-SHIFT controls.

10.4 *Experiment* A7

Obtaining a horizontal trace on an oscilloscope

(a)   Turn the X-SHIFT control until the spot is at the left-hand edge of the screen. Now turn the control so that the spot moves at constant velocity across the screen. When the spot reaches the right-hand edge of the screen, make it 'fly back' to the starting position as quickly as possible. Do this several times — left to right at constant velocity, then right to left as quickly as possible. You are making the spot trace out a horizontal line across the screen.
(b)   You can do the same thing with the Y-SHIFT control to trace out a line in the vertical direction. By manipulating both controls together, the spot can be made to trace out any desired pattern.
(c)   Next switch the TIME/CM (or TIME-BASE) to ON, and set the control to a slow speed, say about 100 ms/cm. You will find that the internal time-base circuitry is doing what you yourself were doing when you rotated the X-SHIFT control. The spot

will be tracing out a horizontal line, moving from left to right at constant velocity and flying back from right to left extremely rapidly.

(d) The speed at which the spot moves from left to right is controlled by the TIME/CM switch. See what happens with the switch at different settings.

(e) When you have obtained a horizontal trace, see how it can be moved up and down and sideways by means of the Y- and X-SHIFTS, and see how it can be moved across the screen by the control marked X-SHIFT.

10.5 *Experiment* A8

Measuring voltages using an oscilloscope

(a) Set the a.c./d.c. switch to d.c. Connect a 9 V battery to the INPUT terminals, as indicated in figure 10.3. Notice the effect on a spot or line trace: the trace moves either up or down, depending upon the battery

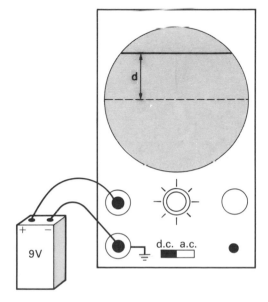

Figure 10.3 Using an oscilloscope to measure d.c. voltage

polarity at the INPUT terminal. By using batteries of different e.m.f.s, you will find that the movement of the spot or line is proportional to the e.m.f. of the battery.

(b) Try out the effects of switching to different VOLT/CM settings and rotating the FINE GAIN control, if one is fitted.

(c) Since the up or down motion of the trace is proportional to the applied voltage, the oscilloscope can be used as a voltmeter. Furthermore, an oscilloscope acts as a very good voltmeter because it has a very high resistance.

(d) In order to use the instrument as a voltmeter, it must be calibrated. This can be done by applying known voltages to the INPUT terminals and observing the movements of the trace. When this is done you should not alter the settings of controls such as INPUT SENSITIVITY, FINE GAIN, etc. You will probably find settings on these knobs marked CAL or VOLTS. When the knobs are in these positions, the VOLTS/CM switch settings should give the voltage calibration directly.

(e) Some instruments have a calibrated Y-SHIFT control. If your scope has one of these, the applied voltage will cause the trace to move up or down. You then rotate the Y-SHIFT control knob to return the trace to its initial position and read off the corresponding voltage from the calibration on the Y-SHIFT scale.

10.6 *Experiment* A9

Showing a sine wave using an oscilloscope

(a) Switch the a.c./d.c. switch to a.c. Obtain a spot on the screen, and hold a finger against the input terminal as shown in figure 10.4. You should observe a vertical trace, the length of which can be varied by means of the VOLTS/CM switch settings. This trace is caused by your body acting as

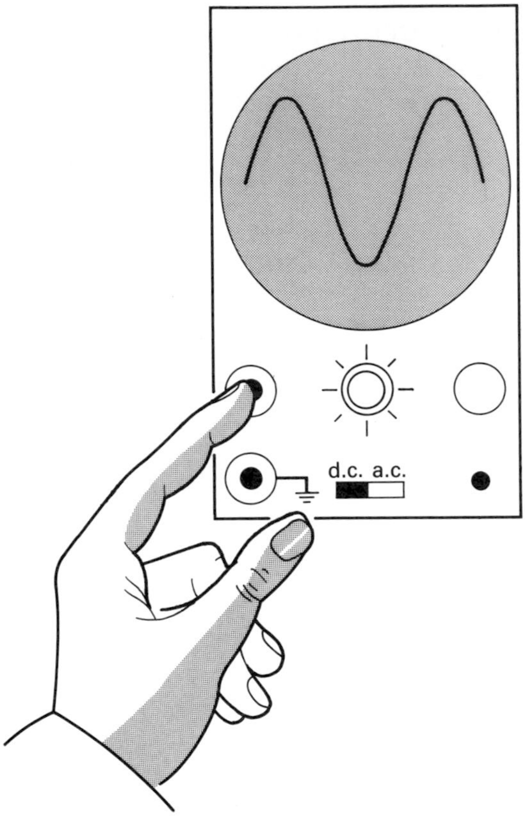

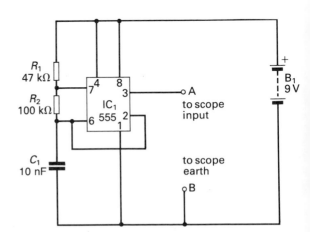

Figure 10.4 Injecting a sine wave

Measuring frequency using an oscilloscope

(a)   Assemble the circuit of figure 10.5 on breadboard as shown in figure 10.6. This circuit is called an *astable multivibrator* and is based on a 555 timer integrated circuit, IC_1. This device is described more fully in Book B. Project Module A6 at the end of this book describes some uses for an astable circuit based on the 555 timer. The two resistors, R_1 and R_2, and the capacitor, C_1, make the circuit produce a high frequency rectangular waveform which can be studied using an oscilloscope.

(b)   Connect the input of the oscilloscope to the wires A and B on the circuit.

(c)   Adjust the VOLTS/CM control and the TIME/CM control so that two complete cycles of the rectangular waveform are displayed as shown in figure 10.7.

(d)   Adjust the X-SHIFT control so that one complete cycle of the rectangular wave can be measured, i.e. between the points C and D in figure 10.7.

(e)   Look at the setting of the TIME/CM control and estimate how long the beam

an aerial to the electromagnetic waves emitted by the power wiring in the room. The waves have a frequency of 50 Hz. They induce a voltage at this frequency into your body and you are coupling a fraction of this voltage into the oscilloscope by means of your finger.

(b)   You can now switch the time-base to ON. This will give you a horizontal trace on which your finger is superimposing a vertical or *Y* deflection of the beam. The combined effect is a sine-wave trace on the screen. If the internal trigger circuits are operating correctly, this should be steady when the TRIGGER switch is set to INTERNAL or AUTO.

Figure 10.5 The circuit of an astable multivibrator

takes to move between the points C and D. This time is known as the *period* of the rectangular waveform. You should find it to be about 1 millisecond (1ms) or 0.001 s.

(f)   The *frequency* of this rectangular waveform is equal to the number of these complete waves in one second. Thus the frequency is given by

$$frequency = \frac{1}{period}$$

$$= \frac{1}{0.001}$$

$$= 1000 \text{ Hz}$$

(g)   The mark-to-space ratio of the rectangular waveform is equal to:

$$\frac{\text{the time the waveform is HIGH}}{\text{the time the waveform is LOW}} = \frac{\text{C to E}}{\text{E to D}},$$

(see figure 10.7)

Can you work out this ratio from the rectangular waveform on the scope?

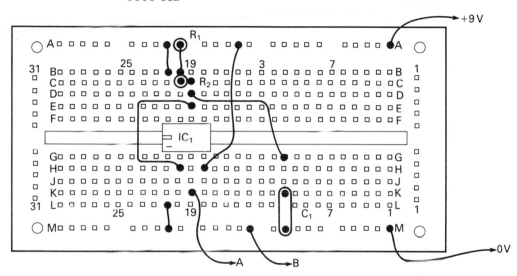

Figure 10.6 The breadboard layout of an astable multivibrator

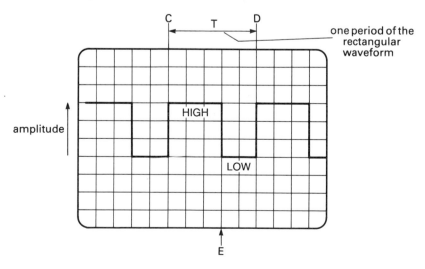

Figure 10.7 A rectangular waveform on the screen of an oscilloscope

11 The Waveform Generator or Signal Generator

▽ 11.1 **What it produces**

A waveform generator is an oscillator which provides a continuously varying voltage across its two output terminals. Figure 11.1 shows what a general-purpose waveform generator looks like. Three waveforms can be selected by push-buttons or a rotary switch: their shapes are *sinusoidal, square,* and *sawtooth* and figure 11.2 shows what they look like.

These waveforms are useful in testing audio and radio circuits. For example, the distortion of a square wave passing through an audio amplifier gives a measure of how high and low frequencies are amplified, known as the *frequency response* of the amplifier. The signals are examined using an oscilloscope.

There are two main types of waveform generator: the *audio frequency (a.f.) generator* (as shown in figure 11.1), and the *radio frequency (r.f.) generator.* Each type has a rotary dial and/or push-buttons so that frequencies over a wide range can be selected. The audio frequency range is from about 30 Hz to 20 kHz but most a.f. generators produce frequencies in the range 0.1 Hz to 100 kHz. The radio frequency generator covers the range from

Figure 11.1 A waveform generator
Courtesy: Farnell International Instruments Ltd.

about 100 kHz to 300 MHz or more. Generally an r.f. generator only produces sine waves, and these waves can usually be *amplitude modulated (AM)* and *frequency modulated (FM)* for use in testing radio receivers.

It is useful to remember ways of writing down different frequencies:

one thousand hertz = 1 kilohertz =
1 kHz = 10^3 Hz
one million hertz = 1 megahertz =
1 MHz = 10^6 Hz
one thousand million hertz = 1 gigahertz
△ = 1 GHz = 10^9 Hz

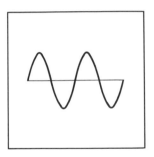 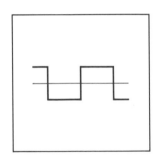 

Figure 11.2 The shapes of waveforms produced by an audio frequency generator

7 11.2 **Using an audio frequency signal generator**

The controls usually consist of the following:

(a) A switch to enable you to choose a particular frequency range.
(b) A knob to tune the instrument to the desired frequency — the large dial in figure 11.1.
(c) An output-voltage control, or VOLUME control. The mains on-off switch might be connected to this control just as on a radio or television set. Sometimes more than one volume control is fitted, in order to make possible coarse and fine control of the output-voltage amplitude. One of these controls might be called an ATTENUATOR and be calibrated in dB (decibels), which is a unit for comparing the ratio of two quantities. When the attenuator control is set to zero, the output voltage is at a maximum. The voltage is reduced as more and more attenuation is switched in.
(d) Output terminals, one of which may be at earth voltage. The terminals may be simple screw types or coaxial-cable sockets. Remember that the outer lead on a coaxial cable is always earthed.

You may find a *resistance* (or to be technically correct, an *impedance*) expressed in ohms marked on the front panel near the output terminals. This is the output resistance (or impedance) of the instrument. Output impedance is the a.c. analogy of the internal resistance of a d.c. battery or power supply. A generator will provide maximum power output when connected to a circuit or *load* equal to its output impedance.

You may find several pairs of output terminals corresponding to different output resistances, e.g. 60 Ω or 600 Ω. Alternatively, there may be only one pair of terminals and a switch to enable you to choose one of several output impedances.
(e) A separate EARTH terminal marked E, $\bumpeq$, may be provided, especially if the instrument does not have a coaxial-cable output socket.

11.3 *Experiment* A11

Using an oscilloscope to check the calibrations on a waveform generator

(a) Set the controls of an audio frequency generator to give a sine wave of about 10 kHz.
(b) Connect the waveform generator to the input of an oscilloscope and obtain a stable trace of two or three complete waveforms on the screen using the trigger control (see Section 10.6).
(c) Use the TIME/CM control to find the period of the waveform (see Section 10.7). Work out the frequency of the waveform and compare your measurement with the calibration on the dial of the generator.
(d) Check the calibration of the waveform generator for other frequencies.

11.4 *Experiment* A12

Producing and using Lissajous figures

(a) Switch off the TIME/CM control and obtain a sharply focused spot of light on the screen of an oscilloscope.
(b) Connect one waveform generator to the X-input of the scope and the other generator to the Y-input as shown in figure 11.3, overleaf. If one of the generator terminals is earthed, make sure that these terminals are connected to the earth terminal on the oscilloscope. The signal to the X-input moves the beam backwards

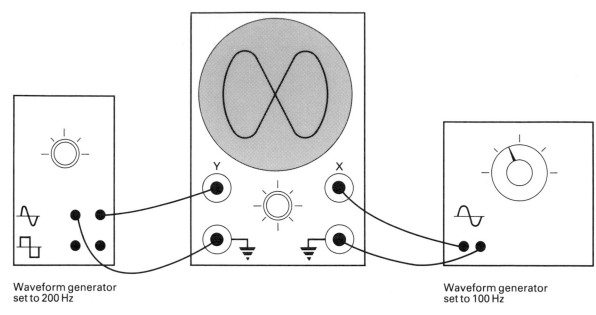

Waveform generator
set to 200 Hz

Waveform generator
set to 100 Hz

Figure 11.3 Making Lissajous figures

and forwards, and the signal to the Y-input moves the beam up and down.

(c)   Set the generator connected to the X-input to give a sine wave of frequency 100 Hz, and the generator connected to the Y-input to give about twice this frequency. Adjust the X and Y sensitivities to give a shape which fills most of the screen.

(d)   Adjust the frequency of the generator connected to the Y-input so that the trace shown in figure 11.3 is obtained. This shape is known as a *Lissajous figure* (pronounced Lease-a-ju). Note that this shape has two loops touching the top of the screen to one loop touching the side.

(e)   Adjust the frequency of the generator connected to the Y-input to 300 Hz,

400 Hz, 500 Hz, etc. and obtain a stable trace in each case. You should find that the number of loops touching the top of the screen will increase from 2 to 5.

(f)   The Lissajous figure is sometimes used to find the frequency of an unknown waveform. In this experiment you will notice that there are two loops touching the top of the screen and one touching the side when the frequencies of the two waveforms are in the ratio 2:1; if three loops touch the top to one loop at the side the frequency ratio is 3:1; and so on.

(g)   What would you expect the pattern to be if the frequencies were in the ratio 3:2? Try it.

12 Project Modules

12.1 **What they are**

At the end of each of the five *Basic Electronics* books there are seven Project Modules for you to build. They are grouped in the books as follows:

Book A
A1: Motor Speed Controller
A2: Pulser
A3: Piezo-sounder and Loudspeaker
A4: Voltmeter
A5: Audio Amplifier
A6: Astable
A7: Triple Lamp

Book B
B1: Schmitt Trigger
B2: Relay Driver
B3: Bistable
B4: Monostable
B5: Radio Receiver
B6: Infrared Source and Sensor
B7: Metal Detector

Book C
C1: Voltage Booster
C2: Geiger Counter
C3: Decade Counter
C4: Thermometer
C5: Bipolar Transistors
C6: Comparator
C7: Strain Meter

Book D
D1: A–D Converter
D2: 8–bit Display
D3: Stepper Motor Driver
D4: Ultrasonic Remote Control
D5: Frequency Divider
D6: Debouncer
D7: Two-Digit Counter

Book E
E1: Logic Gates
E2: BCD Counter
E3: Keyboard Encoder
E4: 4–bit Magnitude Comparator
E5: BCD-Decimal Decoder
E6: Infrared Remote Control
E7: 64–bit Memory

The Project Modules comprise a set of electronic building blocks which are easily interconnected using flying leads to produce useful and interesting electronic systems. Details are given for assembling the Project Modules on printed circuit board (PCB).

A detailed explanation of the function of these do-it-yourself Project Modules is not given. However, frequent references are given to other sections of *Basic Electronics* where a fuller description of the way a component or a circuit works will be found. Ideas for designing electronic systems using the Project Modules are also given. For example, the Pulser (Project Module A2) can be connected to the Piezo-sounder (Project Module A3) to produce alarm signals and audio tones. If the Pulser is coupled with the Relay Driver (Project Module B2), motors and lamps can be switched on and off. And the Loudspeaker (Project Module A3) may be used with the Radio Receiver (Project Module B5) and the Audio Amplifier (Project Module A5) in the design of a portable radio.

Before assembling the do-it-yourself Project Modules, A1 – A7, you should first read Section 6.3 for help on preparing PCBs using the artwork provided. You should also read Section 12.2 which gives hints on handling the CMOS integrated circuit used in Project Module A2.

12.2 **Handling CMOS devices**

Many of the Project Modules and other practical circuits in *Basic Electronics* make use of CMOS integrated circuits. CMOS — pronounced 'seemoss' — stands for Complementary Metal-Oxide Semiconductor and refers to a particular method of making transistors on a silicon chip, as explained in Book C.

Many people worry needlessly about using CMOS integrated circuits and discrete CMOS transistors. It is true that static electricity which sometimes builds up on clothes, carpets, plastics and other electrical insulators, can damage these devices. However, don't worry, for many of today's CMOS devices have special circuits built into them to protect them against damage by static electricity, but it is wise to obey the following simple rules.

(a)   If the CMOS device comes to you wrapped in metal foil or stuck in a piece of plastics foam, leave it there until you are ready to plug it into its IC holder or to solder it into a circuit board

(b)   Make sure that a circuit's power supply is switched off before CMOS integrated circuits are plugged into or pulled out of a circuit board.

12.3 **Identifying resistors and capacitors**

Figure 12.1, and the additional information given with each practical circuit, should help you to identify all the components used in the Project Modules in *Basic Electronics*. However, finding the values of resistors and capacitors may be a problem unless you have worked through Book B. The values of most resistors are indicated by a set of colour bands round them which make up a colour code. Capacitors are made in all shapes and sizes, and some of these, too, have a colour code.

Figure 12.2 (overleaf) shows how to use the colour code for resistors. In general, the fourth band of this code can be ignored. It gives the tolerance of the resistor, and this can be 10% (silver band) or less for the circuits in *Basic Electronics*. Book B explains the meaning of resistance, tolerance and the units of resistance. Note that some manufacturers mark their resistors with five bands: the code is read in the same way except that the third band gives a third digit (for greater precision), and the fourth band is the multiplier band.

Figure 12.3 shows the two main types of capacitor used in the circuits in *Basic Electronics*. Electrolytic capacitors (which generally have values higher than 1 μF) must be connected in circuits the right way round — they are said to be *polarized*. These capacitors are also marked with their *working voltage*. You should make sure that the capacitor you use has a working voltage higher than the d.c. voltage expected across the capacitor otherwise it might break down when the circuit is switched on.

Polyester, and many other types of capacitor, are 'unpolarized' and it doesn't matter which way round they are connected in a circuit. The values of most polyester capacitors (which generally have values less than 1μF) can be read from the colour code marked on them. Their working voltage is usually more than 100 V so they are suitable for all the battery-operated circuits in *Basic Electronics*. Book B explains the properties and use of capacitors more fully.

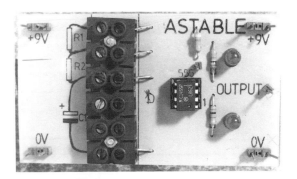

Figure 12.1 The seven project modules described in this chapter

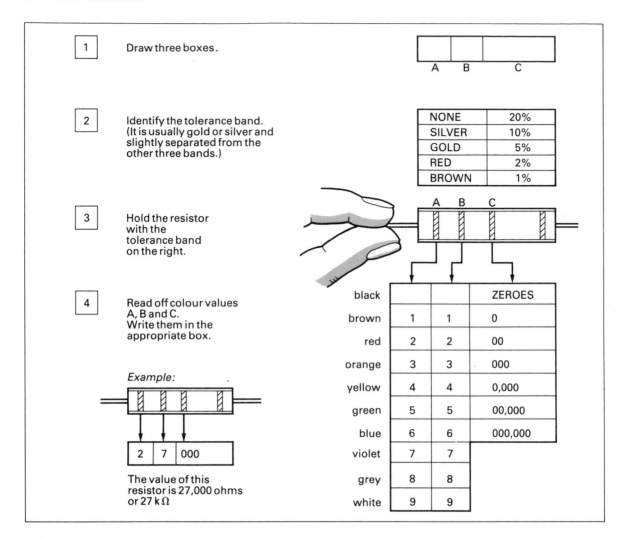

Figure 12.2 Using the resistor colour code

electrolytic		non-electrolytic	
axial (e.g. 4·7 μF, 2·5 V)		A B C (polyester)	
radial (e.g. 10 μF, 16 V)		A–brown, B–black, C–yellow value = 0·1 μF	
		A–brown, B–black, C–orange value = 0·01 μF	
		A–brown, B–black, C–green value = 1 μF	
tantalum (e.g. 2·2 μF, 16 V)		A–red, B–red, C–yellow value = 0·22 μF	
		A–yellow, B–violet, C–yellow value = 0·47 μF	

Figure 12.3 The shapes of capacitors

12.4 *Project Module* A1

Motor Speed Controller

What it does

This Project Module controls the speed of a low-power d.c. motor — the sort used in models and toys — from full speed to full stop simply by adjusting a potentiometer. It may also be used to vary the illumination provided by a low-voltage filament lamp. The Motor Speed Controller is a good beginner's project since it needs just three components.

Circuit

Figure 12.4 shows how an npn transistor, Tr_1, is used as an amplifier to control the current passing through a d.c. motor. The potentiometer, VR_1, is adjusted to control the current flowing into the base terminal, b, of the transistor. The transistor amplifies this small current to produce a much larger current flowing between the collector, c, and emitter, e, terminals of the transistor. This amplified current leaves the emitter terminal and flows through the motor.

This transistor amplifier makes use of negative feedback which makes the motor run at a steady speed even when it has to drive a varying load. Chapter 26 in Book C explains the function of this circuit. Note that the potentiometer, VR_1, controls just the small input current required by the transistor, not all the current required by the motor.

Components and materials

Tr_1: npn transistor, TIP31
VR_1: linear or log variable resistor, value 1 kΩ
R_1: fixed-value resistor, value between 33 Ω and 100 Ω
battery: 9 V PP9, or 5–12 V d.c. power supply
battery clip: PP9 type
wire: multistrand, e.g. 7 × 0.2 mm, for battery and interconnections between other modules
PCB: 90 mm × 50 mm
connectors: PCB plug and socket cut down to make single way connections.

PCB assembly

Figure 12.5 (overleaf) shows the layout of the components on the PCB, and figure 12.6 (overleaf) shows the copper track pattern on the other side of the PCB. See Section 6.3 for guidance on the preparation of the PCB.

Testing and use

Connect a PP9 battery and a d.c. motor, e.g. a Meccano motor, to the module. You will find that the motor's speed can be controlled smoothly using the potentiometer. The Motor Speed Controller is a stand-alone Project Module and is not intended to be connected to other modules described in *Basic Electronics*.

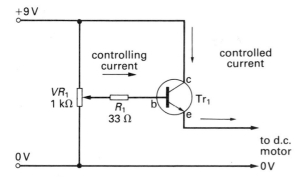

Figure 12.4 Circuit diagram of the Motor Speed Controller

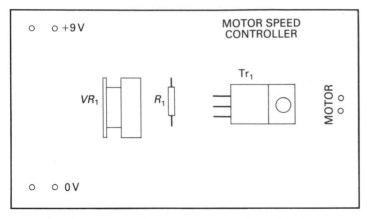

Figure 12.5 Component layout on the PCB (actual size)

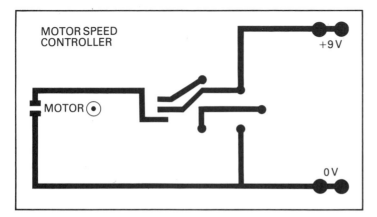

Figure 12.6 Track pattern for the PCB (actual size)

12.5 *Project Module* A2

Pulser

What it does

The Pulser produces a continuous series of on/off signals, i.e. it is a type of oscillator. The frequency of this oscillator can be selected simply by changing the values of two components, a resistor and a capacitor. The on/off signals (or pulses) can be used to control the operation of other circuits, e.g. electronic counters, as explained in Book E of *Basic Electronics*. However, to begin with, it is a simple matter to use the Pulser to produce audio tones as explained below.

Circuit

Figure 12.7 shows how a CMOS integrated circuit, IC_1 is used as the basis of the Pulser. IC_1 contains four independent devices called 2–input NAND four Schmitt triggers, only one of which is used in this circuit. The Schmitt trigger is an important building block in circuits and Project Module B1 describes another use for it.

Resistor R_1, and capacitor C_1, are wired in series so that point X is connected to the input pins 1 and 2, of the Schmitt trigger, IC_1. At the output, pin 3, a light emitting diode, LED_1, flashes on and off to show

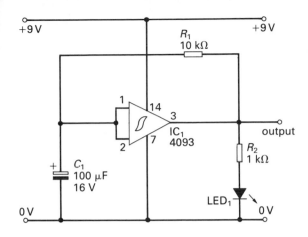

Figure 12.7 Circuit diagram of the Pulser

since $R_1 = 10$ k$\Omega = 10 \times 10^3 = 10^4$ Ω and 100 μF $= 100 \times 10^{-6} = 10^{-4}$ F.

LED$_1$ would pulse at 1 second intervals with these two values. Note that decreasing the value of C_1, or of R_1, raises the frequency: increasing their values, decreases the frequency.

Components and materials

IC$_1$: quad 2–input NAND gate type 4093
R_1, R_2: fixed-value resistors, values 10 kΩ
    and 1 kΩ respectively
LED$_1$: light-emitting diode
C_1: electrolytic capacitor, values e.g.
    100 μF, 220 μF, 16 V working
terminal block: 5-way length
battery: 9V PP9
battery clip: PP9 type
wire: multistrand, e.g. 7 × 0.2 mm, for
    battery and interconnections between
    other modules; single strand
    (1 × 0.6 mm) for wire links
PCB: 90 mm × 50 mm
connectors: PCB plug and socket cut down
    to make single way connections
IC holder: 14-way for IC$_1$
nylon screws: for fixing the terminal block
    to the PCB

PCB assembly

Figure 12.8 shows the layout of the components on the PCB, and figure 12.9 (overleaf) shows the copper track pattern

when the circuit is working. But at frequencies higher than about 20 Hz, the LED will appear to be on continuously. The use of a capacitor and a resistor in electronic timers and oscillators is described in Book B. If you want to calculate the approximate frequency, f, of the pulses produced by the Pulser, use the following equation:

$$f = \frac{1}{(\text{value of } R_1 \times \text{value of } C_1)}$$

In this equation, R_1 must be in ohms and C_1 in farads. Thus, if you select a 10 kΩ resistor and a 100 μF capacitor, the frequency, f, is given by

$$f = \frac{1}{10^4 \times 10^{-4}} = \frac{1}{1} = 1 \text{ Hz}$$

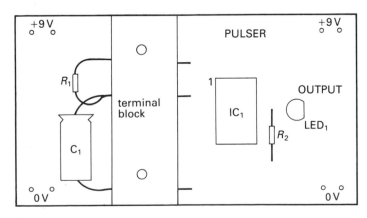

Figure 12.8 Component layout on the PCB (actual size)

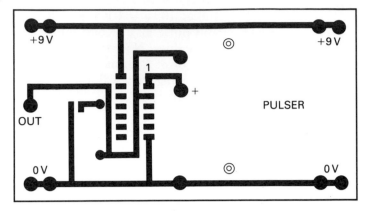

Figure 12.9 Track pattern for the PCB (actual size)

on the other side of the PCB. See Section 6.3 for guidance on the preparation of the PCB.

Make sure that LED$_1$ is connected the right way round. Use 6BA nylon screws to secure the terminal block to the PCB. Note that wire links are needed from the terminal block to the PCB. The terminal pins for power supply and input and output connections are made by cutting single and double pins from a length of PCB connector plug. Similarly slice off single sections of PCB connector socket to make the leads for connecting the Pulser to other Project Modules. Strip 5 mm of insulation from the ends of wires, and use a crimping tool to squeeze a crimp connector on the bare ends. Push the crimp connector into the PCB socket until it clicks into place.

Testing and use

Use a small screwdriver to fix R_1 (10 kΩ) and C_1 (100 μF) to the terminal block. Connect the 9 V battery and LED$_1$ will begin to flash at intervals of about 1 s. Note that the equation above only approximately determines the frequency. If you want a precise frequency of, say, 1 Hz, R_1 should be replaced by a 22 kΩ variable resistor.

Replace R_1 by a light-dependent resistor (LDR), and see what happens to the frequency of the pulses as you cover and uncover the LDR. Can you work out how the resistance of the LDR varies with the amount of light reaching it? The properties and use of the LDR are described in Book B, further use for the Pulser is described in Project Module A3.

12.6 *Project Module* A3

Piezo-sounder and Loudspeaker

What they do

They provide a sound output when fed with varying elecrical signals generated by other Project Modules (e.g. the Pulser, Project Module A2). The Piezo-sounder provides a low-power audio output, e.g. a buzz in alarm circuits and a click in counting circuits. And the Loudspeaker provides a higher power audio output, e.g. music and speech, in circuits such as the Radio Receiver (Project Module B5).

The Piezo-sounder requires very little electrical power so it can be operated direct from the output of CMOS integrated circuits, e.g. the Pulser (Project Module A2). The Piezo-sounder is based on a ceramic material which changes its shape when a voltage is applied across it. Changes of voltage cause changes of shape which, in turn, generates a sound which follows the pattern of voltage changes. Conversely, though not the way the Piezo-sounder is used here, the ceramic material will generate a voltage when it is squeezed, a property called piezoelectricity and put to good use in some forms of gas lighter.

The Loudspeaker requires more power to operate it since a varying current is fed to a coil which is free to move in a strong magnetic field and attached to the cone of the Loudspeaker. Changes of current produce (electromagnetic) forces which act on the coil so that the cone moves and generates sounds that 'follow' the pattern of current changes.

Components and materials

piezo-sounder: about the size of 10p coin with a hole in the middle with two short lengths of wire from it

loudspeaker: a low-power type about 60 mm in diameter with two solder tags on it: its resistance should not be less than 35 Ω

wire: multistrand, e.g. 7 × 0.2 mm, for connections to the outputs of other Project Modules

PCB: 90 mm × 50 mm for the Piezo-sounder

connectors: PCB plug and socket cut down to make single way connections

Printed circuit board assembly

Figure 12.10 shows the simple PCB mounting of the Piezosounder, and figure 12.11 (overleaf) the copper track pattern on the other side of the PCB.

The terminal pins for the power supply and Piezo-sounder connections are made by cutting single and double pins from a length of PCB connector plug. Similarly, slice off single pin sections of PCB socket to make up leads for connecting the Piezo-sounder to other Project Modules. Strip 5 mm of insulation from the ends of wires, and use a crimping tool to squeeze a crimp connector on the bare ends. Push the crimp connector into the PCB socket until it clicks into place.

The Loudspeaker is mounted in a flat cylinder of rigid material (e.g. the screw cap of a coffee jar) to protect its fragile paper cone from damage, and to create a baffle which enhances the sound output — see figure 12.12, overleaf. Leads at least one metre long should be soldered to the loudspeaker and terminated with PCB connector sockets. Make sure that the leads are anchored so that they do not tug on the solder tags.

Testing and use

The Piezo-sounder is easily tested by connecting it to the Pulser (Project Module

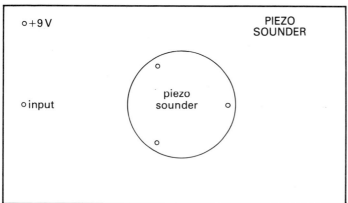

Figure 12.10 Mounting of Piezoelectric-sounder to the PCB

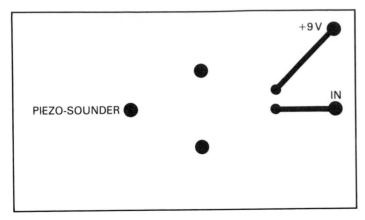

Figure 12.11 Track pattern for the Piezo-sounders (actual size)

A2) as shown in figure 12.13, overleaf. Make sure the Pulser produces pulses by using a value of 10 kΩ for R_1 and a value of 100 μF for C_1 (see figure 12.6). A regular clicking sound will be heard from the Piezo-sounder in time with the flashing of LED_1. Change the value of C_1 to 100 nF (1000 times smaller than 100 μF), and a high-pitched tone will be heard.

Use a value of 100 nF for C_1 in the Pulser,

and replace R_1 by a light dependent resistor (LDR). The LDR is described in Chapter 7, Book B. By varying the illumination of the LDR, you will be able to vary the frequency of the audio tone.

How does the resistance of the LDR vary with the amount of light falling on it?

The Pulser does not produce enough power to operate the Loudspeaker, and it is best tested by connecting it to the output of the Audio Amplifier (Project Module A5). However, if you have two loudspeakers, try this: connect the two loudspeakers together by long wires and send a friend with one of them into another room. Close the door between the rooms and see if you can communicate with each other using the loudspeakers as both earpieces and microphones. A much better intercom than this one requires the use of the Audio Amplifier (Project Module A5).

Where does the electrical power come from to make this simple intercom work?

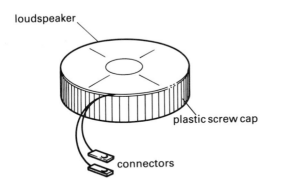

Figure 12.12 Wiring for the Loudspeaker

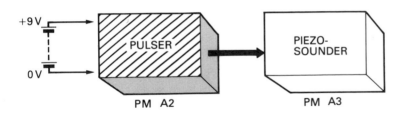

Figure 12.13 Testing the Piezo-sounder

12.7 *Project Module* A4

Voltmeter

What it does

This Project Module measures potential difference on three d.c. ranges, 1 V, 10 V and 100 V full scale deflection (f.s.d.). It is intended to drive a 1 mA f.s.d. moving coil meter, or the 1 mA range of a multimeter. The Voltmeter will effectively give these meters a 2.5 MΩ input resistance for measuring voltage, and it therefore provides a useful upgrading of cheap multimeters.

Circuit

The circuit shown in figure 12.14 uses an operational amplifier (op amp), IC_1, which has a high input impedance. In this circuit, IC_1 is operated with 100% negative feedback since the output, pin 6, is shorted to the inverting input, pin 2. Thus, in this circuit, the op amp does not act as an amplifier but as a high-to-low-impedance buffer. This arrangement allows the high-value resistors R_1 to R_4 to determine the resistance of the voltmeter which therefore

has an overall resistance of about 2.5 MΩ on all three ranges. Thus the circuit draws very little current from the circuit under test and more accurate readings are obtained with this voltmeter than with cheap multimeters which might have a resistance of as low as 2 kΩ. The way the resistance of a voltmeter determines the accuracy of readings measured is described in Section 9.3. Operational amplifiers are discussed fully in Book D.

The Voltmeter is designed to operate a 1mA f.s.d. moving coil meter, or the 1mA range on a multimeter. Thus 1 V d.c. across the input terminals of the Voltmeter gives a full scale deflection reading on the 1 mA meter. Resistors R_1 to R_4 comprise a voltage divider which attenuates (reduces) the input voltage allowing the rotary switch, SW_1, to select the ranges, 1 V, 10 V and 100 V f.s.d. on the meter.

Transistor Tr_1 and resistors R_5 to R_6 protect the meter from overload, e.g. if 10 V is applied when SW_1 is set to 1 V f.s.d. They also protect IC_1 which should not have more than 8 V difference between

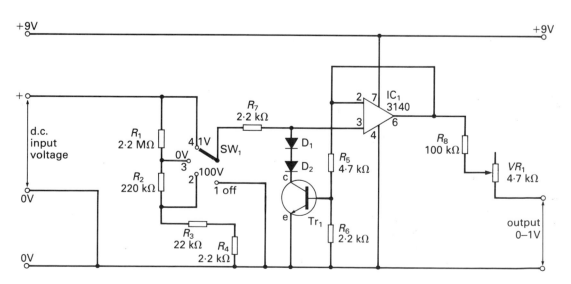

Figure 12.14 Circuit diagram of the Voltmeter

its input pins 2 and 3. As soon as the output voltage at pin 6 exceeds about 1.3 V, Tr_1 switches on, keeping the voltage on pin 3 below about 1.3 V as current flows through the diodes D_1 and D_2 and through the collector-emitter terminals of Tr_1. Note that this overload protection begins to operate when the current to the meter exceeds about 30% of its f.s.d., but this excess does no harm to the meter.

Components and materials

IC_1: operational amplifier, type 3140
Tr_1: bipolar transistor, type BC108 or
 similar
IC holder: 8-way
R_1 to R_8: fixed-value resistors, 0.25 W, ±
 5%
D_1, D_2: silicon diodes, type 1N4001 or
 similar

SW_1: 2-pole 6-way, or 3-pole 4-way
VR_1: miniature preset trimmer resistor,
 value 4.7 kΩ
battery: 9 V PP9
battery clip: PP9 type
meter: 1 mA f.s.d. moving coil type, or
 1 mA range on multimeter
wire: multistrand, e.g. 7 × 0.2 mm.
PCB: 90 mm × 50 mm
connectors: PCB header and PCB socket
 housing; crimp terminals

PCB assembly

Figure 12.15 shows the layout of the components on the PCB, and figure 12.16 shows the copper track pattern on the other side of the PCB. See Section 6.3 for guidance on the preparation of the PCB.

 Cut off all but one of the poles, and four consecutive tags, numbered 1 to 4, round

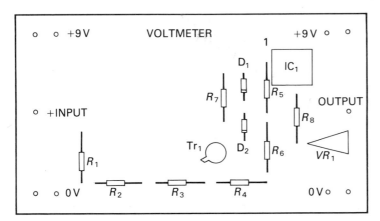

Figure 12.15 Component layout on the PCB (actual size)

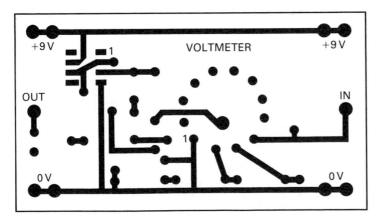

Figure 12.16 Track pattern for the PCB (actual size)

the edge of SW_1. Cut off the circular ends of the remaining tags to leave five short terminal pins on the switch. These pins are to pass through the holes on the PCB.

The terminal pins for the connections on the PCB are made by cutting single and double pins from the PCB header. Single sections of the PCB socket housing are used for terminating wire ends for connecting the Voltmeter to the battery and the 1 mA meter. Strip 5 mm of insulation from the ends of the wires, and use a crimping tool to squeeze a crimp connector on the bare ends. Push the crimp connector into the PCB socket until it clicks into place.

Testing and use

Measure the e.m.f. of a 9 V battery using a voltmeter you know is accurate. Connect a 9 V battery to the Voltmeter, and switch SW_1 to its 10 V range which is the third switch position clockwise from the 'off' position. Connect the battery of known e.m.f. to the input terminals of the circuit making sure that the polarity is correct. Now adjust VR_1 so that the meter reads the measured e.m.f. of the calibration battery. Now that you have calibrated the circuit for the 10 V range, the other ranges will automatically be correct because of the values selected for resistors R_1 to R_4.

12.8 *Project Module* A5

Audio Amplifier

What it does

This Project Module amplifies audio frequency signals, e.g. those from a microphone, and delivers enough power to operate a small loudspeaker. The Audio Amplifier can be used to boost the sound from the Radio Receiver (Project Module B5). It can also be used in an intercom, or with the Infrared Source and Sensor (Project Module B6) in an optical communications system.

Circuit

The circuit shown in figure 12.17 uses an operational amplifier (op amp), IC_1, as a preamplifier. The strength of the output signal from the op amp is varied with the aid of the variable resistor, VR_1, which

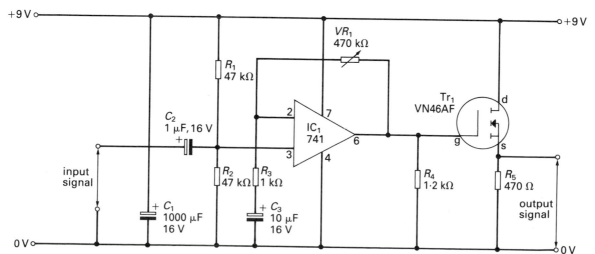

Figure 12.17 Circuit diagram of the Audio Amplifier

therefore acts as a volume control. This signal is fed to transistor Tr_1 which acts as a power amplifier to drive the loudspeaker. The large-value electrolytic capacitor, C_1, stabilizes the operation of the amplifier, and capacitor C_2 is a coupling capacitor which transfers the signal from the source of signals to pin 3 of the op amp. Capacitors are discussed in Book B, the op amp, IC_1, in Book D and the field-effect transistor, Tr_1, in Book C.

Components and materials

IC_1: operational amplifier, type 3140
VR_1: linear or log variable resistor, value 470 kΩ
R_1 to R_5: fixed-value resistors, 0.25 W, $\pm5\%$
IC holder: 8-way

Tr_1: n-channel VMOS field-effect transistor, type VN46AF
C_1 to C_3: electrolytic capacitors, 1000 μF, 1 μF and 10 μF, respectively, 16 V working
battery: 9 V PP9
battery clip: PP9 type
wire: multistrand, e.g. 7 $\times$ 0.2 mm
PCB: 90 mm $\times$ 50 mm
connectors: PCB header and PCB socket housing; crimp terminals
bolt: 1 $\times$ 6BA (for fixing transistor Tr_1)

PCB assembly

Figure 12.18 shows the layout of the components on the PCB, and figure 12.19 shows the copper track pattern on the other side of the PCB. See Section 6.3 for guidance on the preparation of the PCB.

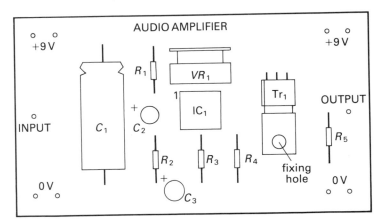

Figure 12.18 Component layout on the PCB (actual size)

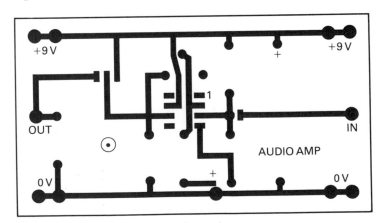

Figure 12.19 Track pattern for the PCB (actual size)

The terminal pins for the connections on the PCB are made by cutting single and double pins from the PCB header. Single sections of the PCB socket housing are used for terminating wire ends for connecting the Audio Amplifier to the battery and the Loudspeaker (Project Module A3). Strip 5 mm of insulation from the ends of the wires, and use a crimping tool to squeeze a crimp connector on the bare ends. Push the crimp connector into the PCB socket until it clicks into place.

Testing and use

Make the Pulser (Project Module A2) produce an audio tone of about 1 kHz by choosing a value of 100 nF for C_1 and a value of 10 kΩ for R_1. Connect the 9 V and 0 V power supply pins between the two boards, and the output of the Pulser to the input of the Audio Amplifier. Connect the 9 V battery and the Loudspeaker (Project Module A3) to the Audio Amplifier. Vary the volume control, VR_1, so that you can control the output power delivered to the loudspeaker.

Alternatively, if you have two loudspeakers, use one as a microphone connected to the input, and the other connected to the output of the Audio Amplifier. Adjust VR_1 to give maximum amplification and place the two loudspeakers close to each other. A 'howl' should be heard from one of the loudspeakers. Now separate the two loudspeakers and use the circuit as a one-way intercom to talk between two rooms.

How would you use two single-pole, double-throw switches (see Chapter 5) in this intercom so that each loudspeaker can be used as a microphone or as a loudspeaker depending on whether you are talking or listening?

Figure 12.20 shows how to use the Audio Amplifier in an AM radio. You will need Project Module B5 in addition to the Loudspeaker (A3) for this design. See Project Module B6 for further applications of the Audio Amplifier.

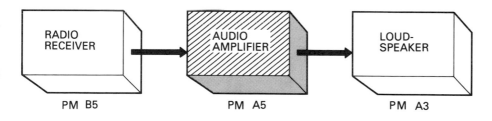

Figure 12.20 System application: AM Radio

12.9 *Project Module* A6

Astable

What it does

This Astable (or astable multivibrator to give it its full title) produces a continuous series of on/off pulses. Like the Pulser, it is an oscillator but it has two advantages over the Pulser. First, the on and off (high and low) times of the pulses can be individually 'programmed' by changing the values of two timing resistors. The Pulser merely produces a series of square waves, having equal on and off times. Second, the Astable provides more output power than the Pulser so that it can drive a relay or loudspeaker direct from its output. The Astable is useful in digital counting and control circuits.

Circuit

The circuit shown in figure 12.21 is similar to the one described in Experiment A10 and designed to provide signals for display on an oscilloscope. The Astable is based on an integrated circuit, IC_1, the 555 timer. This device is described in Chapter 14, Book B and Chapter 9, Book D. The frequency of the pulses generated by the 555 timer is determined by just three components, R_1, R_2 and C_1. In this circuit the pulses switch two light-emitting diodes, LED_1 and LED_2, on and off alternately which is helpful for seeing if the circuit is working.

As explained in Chapter 14, Book B, there are two simple equations which give, approximately, the times for which each LED is on.

LED_2 is on for a time equal to
$$t_1 = (R_1 + R_2) \times C_1 \text{ seconds}$$
LED_1 is on for a time equal to
$$t_2 = R_2 \times C_1 \text{ seconds}$$
These times are approximate because the values of resistors and capacitors differ from the values marked on them due to unavoidable manufacturing tolerances. The values of the resistors must be in ohms and the value of the capacitor in farads. Thus, for the values shown in figure 12.20,
$R_1 = 50 \text{ k}\Omega = 5 \times 10^4 \text{ ohm}$,
$R_2 = 100 \text{ k}\Omega = 10^5 \text{ ohm}$, and
$C_1 = 10 \text{ μF} = 10 \times 10^{-6} = 10^{-5} \text{ farads}$.

Therefore $t_1 = 1.5 \times 10^5 \times 10^{-5} = 1.5 \text{ s}$ and
$$t_2 = 10^5 \times 10^{-5} = 1 \text{ s}$$

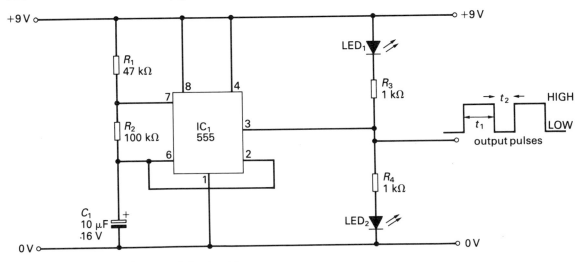

Figure 12.21 Circuit diagram of the Astable

Thus LED$_2$ is on for 1.5 s and LED$_1$ is on for 1 s. Clearly you can easily change these times. Thus if the value of C_1 is changed to 100 μF, t_1 and t_2 will increase 10 times to 15 s and 10 s, respectively. A smaller value for C_1, e.g. to 10 nF, would reduce t_1 and t_2 by a factor of 1000. Note that it is not possible to change the value of R_2 without altering both t_1 and t_2, whereas altering R_1 just changes the value of t_1.

The frequency, f, of the pulses produced by the Astable is given by

$$f = 1/(t_1+t_2) \text{ Hz}$$

Thus if $C_1 = 10$ nF, $R_1 = 50$ kΩ and $R_2 = 100$ kΩ,

$$f = 400 \text{ Hz}$$

Components and materials

IC$_1$: integrated circuit timer, type 555
R_1 to R_4: fixed-value resistors, values
47 kΩ, 100 kΩ, 1 kΩ, 1 kΩ, respectively
0.25 W, ±5%
IC holder: 8-way
LED$_1$, LED$_2$: light-emitting diodes
C_1: electrolytic capacitor, 10 μF 16 V working
terminal block: 5-way length
battery: 9 V PP9
wire: multistrand, e.g. 7 × 0.2 mm; single strand, e.g. 1 × 0.6 mm
PCB: 90 mm × 50 mm
connectors: PCB header and PCB socket housing; crimp terminals.

PCB assembly

Figure 12.22 shows the layout of the components on the PCB, and figure 12.23 (overleaf) the copper track pattern on the other side of the PCB. See Section 6.3 for guidance on the preparation of the PCB.

Make sure that LED$_1$ and LED$_2$ are connected the right way round. The cathode of LED$_1$ faces away from the edge of the board, and the cathode of LED$_2$ towards the edge of the board. Use nylon screws to secure the terminal block to the PCB. Note that four wire links are needed between the terminal block and the PCB.

The terminal pins for the connections on the PCB are made by cutting single and double pins from the PCB header. Single sections of the PCB socket housing are used for terminating wire ends for connecting the Astable to the battery and to other Project Modules. Strip 5 mm of insulation from the ends of the wires, and use a crimping tool to squeeze a crimp connector on the bare ends. Push the crimp connector into the PCB socket until it clicks into place.

Testing and use

Use a small screwdriver to connect R_1, R_2 and C_1, shown in figure 12.22, to the terminal block. Connect the 9 V battery to the circuit and check that LED$_1$ and LED$_2$

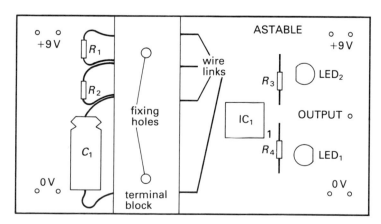

Figure 12.22 Component layout on the PCB (actual size)

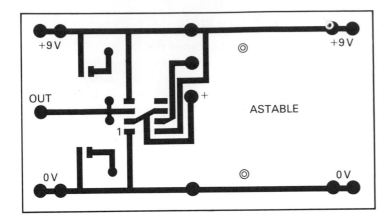

Figure 12.23 Track pattern for the PCB (actual size)

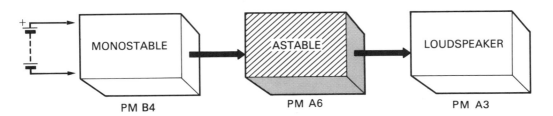

Figure 12.24 System application: Timer Alarm

flash on and off alternately. Change the value of C_1 to 100 μF and see whether the times for which LED_1 and LED_2 are lit has increased by ten times. Connect the Loudspeaker (Project Module A3) to the Astable and note the clicks produced by the loudspeaker. Replace C_1 by a 10 nF capacitor and note the 400 Hz audio tone

produced by the loudspeaker. When you have built the Relay Driver (Project Module B2) the Astable can be used for precise on/off control of d.c. motors.

For the application shown in figure 12.24 you will need the Monostable (Project Module B4). It provides an alarm after a time delay set by the Monostable.

12.10 *Project Module* A7

Triple Lamp

What it does

Its red, amber and blue lamps are switched on and off electronically, using transistors which respond to an appropriate input signal. These lamps offer a brighter display in systems design than light emitting

diodes, and their different colours enable you to code light signals as in pedestrian crossings and other warning systems. The lamps can be driven on and off by the output signals from a number of other Project Modules, including Logic Gates (Project Module E1), Schmitt Trigger (Project Module B1) and Astable (Project Module A6).

Circuit

Figure 12.25 shows how three Darlington pair transistors, Tr_1 to Tr_3, are used to switch the lamps on and off. In Chapter 20 of Book C you can read how the Darlington pair works. It is very sensitive to a signal applied to the inputs of the transistors so it is an ideal device for 'interfacing' between power-hungry devices such as lamps and low-power devices such as CMOS integrated circuits. So the Pulser (Project Module A2) can drive the Triple Lamp module.

To prevent damage to the transistors, the circuit incorporates three protection resistors, R_1 to R_3. Output pins enable the module to be used to operate other Project Modules if required. A minimum positive voltage of about 1.4 V is required at any input to switch on the relevant lamp. For voltages less than this, the lamps are off. The input current required to switch on a lamp is less than 0.1 mA. Depending on the current gain of the transistors used, the current might be less than 0.01 mA (10 μA).

Components and materials

Tr_1 to Tr_3: Darlington pair, type ZTX600
R_1 to R_3: fixed-value   resistors, values
  2.2 kΩ, 0.25 W, ±5%

L_1 to L_3: LES lampholders, 'flat top type';
  6 V, 60 mA LES lamps
battery: 9 V, PP9
wire: multistrand, e.s. 7 × 0.2 mm; single
  strand, e.g. 1 × 0.6 mm
PCB: 90 mm × 50 mm
connectors: PCB header and PCB socket
  housing; crimp terminals

PCB assembly

Figure 12.26 (overleaf) shows the layout of the components on the PCB, and figure 12.27 (overleaf) the copper track pattern on the other side of the PCB. See Section 6.3 for guidance on the preparation of the PCB. Check that the transistors are connected the right way round in the PCB. The terminal pins for the connections in the PCB are made by cutting single and double pins from the PCB header. Single sections of the PCB socket housing are used for terminating wire ends for connecting the Triple Lamp to the battery and to other Project Modules. Strip about 5 mm of insulation from the ends of the wires, and use a crimping tool to squeeze a crimp connector to the bare ends. Push the crimp connector into the PCB socket until it clicks into place.

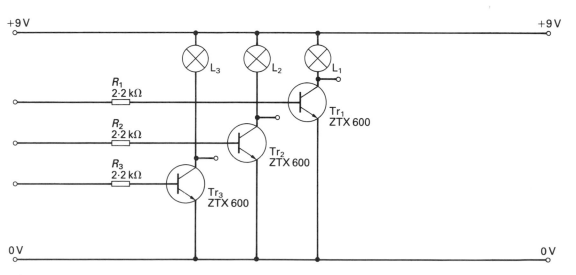

Figure 12.25 Circuit diagram of Triple Lamp

Testing and use

Connect a 9 V battery to the module and the lamps should remain off. Now bridge your finger across one of the inputs and the +9 V power supply pin. The relevant lamp should light. Test the other inputs in the same way. The lamps should light if you use your body as the conducting path for the input signal. Try and set a lamp to light by making several people link hands while the first and last in the chain complete the circuit via an input pin and +9 V.

Now wire up the Pulser (Project Module A2) so that it flashes its LED at a fairly slow rate. Interconnect the power supply pins of the Pulser and Triple Lamp modules. Connect the output of the Pulser to one of the inputs of the Triple Lamp so that the relevant lamp flashes. Replace the Pulser with the Astable (Project Module A6). Use the Triple Lamp in the systems you design with the Project Modules to provide a brighter output signal than the built-in LEDs give.

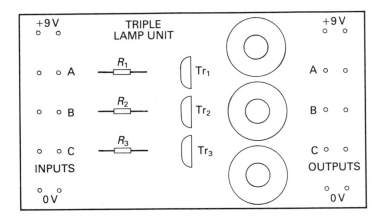

Figure 12.26 Layout of components on the PCB

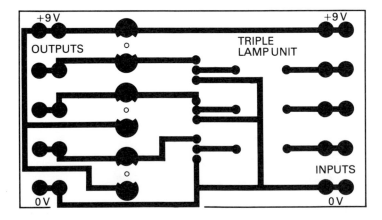

Figure 12.27 Copper track pattern of PCB

13 Questions and answers — Book A

Revisions questions

General

1 Name six objects in the home which use electronic devices.

2 What is a telecommunications system?

3 What is the integrated circuit called at the heart of a microcomputer?

4 What is software?

5 Describe some ways electronics helps doctors and surgeons look after their patients.

6 Draw a block diagram showing the function of a thermometer.

7 Give an example of a control system.

8 Name one function of your body which can be described as digital.

9 How are analogue and digital watches usually distinguished from each other?

10 Why is the temperature of the air around us an analogue quantity?

11 Electrons flow from the negative terminal of a battery to its positive terminal.
   True or false?

Circuits

12 When two lamps are connected in series, is the current through one lamp different from or equal to the current through the other lamp?

13 The total current flowing through two lamps connected in parallel is the difference between the currents flowing through each lamp. True or false?

14 Draw a circuit containing a battery and three lamps connected in series with each other.

15 Look at the circuit shown in figure 13.1. How would you describe this circuit? Which lamp is brightest? Why?

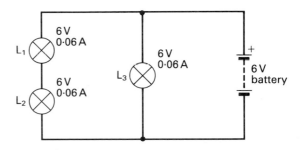

Figure 13.1

Switches

16 Name three types of mechanical switch.

17 A reed switch is operated by...
   heat, noise or magnetism?

18 Name three uses for a reed switch.

19 Two identical switches are needed to switch a motor on and off from two different positions. What type of switch is required: (a) a single-pole, double-throw, or (b) a double-pole, double-throw?

20 Two students each have a switch. Draw a circuit diagram which will enable a warning lamp to flash on when either switch is pressed.

21 What changes would you make to the circuit of question 20 if the light is to come on only when both students press their switch?

22 A rocket being prepared for launching has two people in it, each of whom has a two-position switch, one position marked 'hold' and one marked 'go'. Draw a circuit so that:
   (a) a 'hold' light comes on when either person has their switch in the 'hold' position, and
   (b) a 'go' light comes on only if both people have their switch in the 'go' position.

Atomic structure

23 Electrons are present in all materials on earth.
   True or false?

24 What is happening to an atom which is radioactive?

25 The nucleus of an atom is a very small part of the space in an atom.
   True or false?

26 Electrons are held in shells of an atom by a force which exists between these electrons and the protons in the nucleus. Is this force *magnetic* or *electrical*?

27 Does a hydrogen atom have one, two or no protons in its nucleus?

28 How does deuterium differ from hydrogen? How is it similar to hydrogen?

29 A water molecule is produced when two atoms of hydrogen combine with one atom of oxygen. How many protons are there in a molecule of
   (a) ordinary water (b) heavy water?

30 A carbon-12 atom has six protons in its nucleus. How many electrons and neutrons make up a normal carbon-12 atom?

31 Electrons can easily be removed from the atoms of materials which are electrical conductors.
   True or false?

32 When electrons are removed from the shells of an atom, the atom is said to be in an . . .
   unfortunate, light or ionised state?

Meters

33 A moving coil meter makes use of a magnetic field produced by a flow of electrons.
   True or false?

34 The maximum reading on a meter is known as......

35 A voltmeter is used for measuring potential difference.
   True or false?

36 An ammeter measures *electric current*, *resistance* or *voltage*?

37 Voltmeters are placed in.......with a component across which voltage is being measured in a circuit.
   Series or parallel?

38 A moving coil ammeter should have a.....resistance.
   High, low or medium?

39 The best voltmeters require a high current to operate them.
   True or false?

40 The ohms-per-volt rating of a voltmeter is equal to . . .
   Full-scale voltage, internal meter resistance or reciprocal of the full-scale current?

41 A voltmeter uses a meter movement with a 100 μA f.s.d. What is the ohms-per-volt rating of the instrument?
   320 Ω, 10 kΩ, 25 kΩ?

Cathode-ray oscilloscopes

42   Cathode rays are...
rapidly moving molecules, a form of laser beam or a beam of high-speed electrons?

43   What is the purpose of an electron gun in an oscilloscope?

44   An oscilloscope displays the waveform shown in figure 13.2. If the Y-sensitivity control is set at 2 V/cm, what is the amplitude of this waveform?

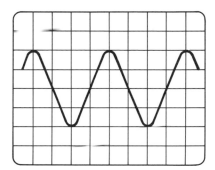

Figure 13.2

45   An oscilloscope displays the waveform shown in figure 13.3. If the TIME/CM is set at 2 ms/division, what is the period and frequency of the wave?

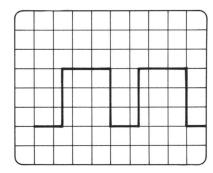

Figure 13.3

46   What is the mark-to-space ratio of the waveform shown in figure 13.3?

47   The four diagrams in figure 13.4 show the waveforms of signals displayed on an oscilloscope. Which figure shows

(a)   an a.c. supply
(b)   audio signals from a microphone
(c)   a varying signal when the time-base of the oscilloscope is turned off
(d)   a signal from an astable multivibrator?

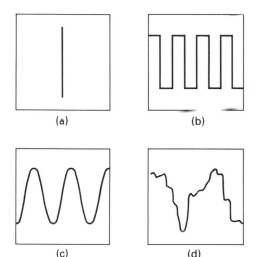

Figure 13.4

Revision answers

29   (a) 10   (b) 10
30   6 electrons, 6 neutrons
41   10 kΩ/V
44   4 V
45   8 ms; 125 Hz
46   1.67

Answers to questions — Book A

Section 1.7
8   (c)
9   (b)
10   (b)
Section 3.6
1   (b)
6   (a), (c), (d)

Section 5.5
1

A	B	C	Lamp
0	0	0	0 (off)
1	0	0	0
0	1	0	0
0	0	1	0
1	1	0	1 (on)
1	0	1	1
0	1	1	0
1	1	1	1

3

SW_1	SW_2	Lamp
1	0	1 (on)
1	1	0 (off)
0	1	1
0	0	0

Section 7.3
1

Oxygen–18	8	8	10	18
Copper–63	29	29	34	63
Silver–108	47	47	61	108
Silicon–28	14	14	14	28
Germanium–74	32	32	42	74
Carbon–12	6	6	6	12
Carbon–14	6	6	8	14
Iron–56	26	26	30	56

Section 8.3
2   0.000 005 s or 5 μs
Section 8.4
1   2.1 A